A FAMILY VENTURE

A FAMILY VENTURE

MEN AND WOMEN ON THE SOUTHERN FRONTIER

JOAN E. CASHIN

THE JOHNS HOPKINS UNIVERSITY PRESS
Baltimore and London

Printed in the United States of America on acid-free paper

Originally published in a hardcover edition by Oxford University Press, Inc., 1991
Johns Hopkins Paperbacks edition, 1994
03 02 01 00 99 98 97 96 95 94 5 4 3 2 1

The Johns Hopkins University Press
2715 North Charles Street
Baltimore, Maryland 21218-4319
The Johns Hopkins Press Ltd., London

Portions of Chapter 1 first appeared in Joan E. Cashin, "The Structure of Antebellum Planter Families: 'The Ties That Bound Us Was Strong,'" *Journal of Southern History* 56 (February 1990): 55-70

Library of Congress Cataloging-in-Publication Data

Cashin, Joan E.
A family venture : men and women on the southern frontier / Joan E. Cashin.
p. cm.
Originally published : New York : Oxford University Press, 1991.
Includes bibliographical references and index.
ISBN 0-8018-4964-0 (pbk. : alk. paper)
1. Plantation life — Southern States — History — 19th century. 2. Plantation owners — Southern States — History — 19th century. 3. Migration, Internal — Southern States — History — 19th century. 4. Frontier and pioneer life — Southern States. 5. — Southern States — History — 1775-1865. I. Title.
[F213.C34 1994]
306′.0975—dc20 94-14544
CIP

A catalog record for this book is available from the British Library.

For my grandparents

Acknowledgments

It is a pleasure to acknowledge the kind assistance I have received from many people while writing this book. Financial aid from an Albert J. Beveridge Grant from the American Historical Association, a Faculty Research Fellowship and a Special Research Grant from Southern Illinois University, the Charles Warren Center at Harvard University, the Whiting Foundation, and the Danforth Foundation supported the research. The staffs at Widener and Lamont Libraries at Harvard University, Morris Library at Southern Illinois University, and Robeson Library of Rutgers University at Camden arranged loans of many books and reels of microfilm.

Many thanks must also go to the Library of Congress; the Virginia Historical Society; the Alderman Library at the University of Virginia; the Southern Historical Collection at the University of North Carolina; the South Caroliniana Library at the University of South Carolina; the Barker Texas History Center at the University of Texas at Austin; and the Departments of Archives and History in the states of Virginia, North Carolina, South Carolina, Alabama, Mississippi, and Texas for their assistance during my research travels. Mimi Rogers of the Alabama Department of Archives and History, Richard Shrader of the Southern Historical Collection, and Allen Stokes of the South Caroliniana Library were especially helpful. Henry Fulmer of the South Caroliniana Library deserves a special tribute for his graciousness and good humor. Conversations along the way with colleagues at Harvard University, Southern Illinois University, and Rutgers University enriched my thinking about the project. Vernon Burton and Richard Dunn generously shared with me their knowledge of several Southern families.

I am much indebted to the teachers, colleagues, and friends who read the manuscript on its eventful journey from dissertation to book, beginning with David Herbert Donald. When he gave a meticulous reading to the dissertation, I gained a great deal from his knowledge of many aspects of Southern history; I have always been inspired by his example as a scholar. Edward Ayers read the dissertation and significant portions of the revised manuscript, offering shrewd criticisms in a kind, collegial spirit. Kenneth Greenberg, Reginald Horsman, Larry Schweikart, Joel Williamson, and Jacob Weiner read sections of the manuscript pertinent to their areas of expertise and made constructive criticisms.

I am particularly grateful to Catherine Clinton, J. William Harris, Alan Kraut, and David Rankin, who read the entire manuscript when it was nearly done. Their comments on matters of style and substance were astute and invariably thought-provoking; I did not always agree with their suggestions, but I appreciated them all. Rachel Toor at Oxford University Press convinced me to make one last round of revisions, and India Cooper copyedited the manuscript with great care; Karen Wolny capably supervised the book's production. All errors or infelicities of style are my responsibility, of course.

This book is dedicated with love to the four individuals who introduced me to the mysteries of time and place.

Camden, N.J. J. E. C.
January 1991

Contents

A FAMILY VENTURE

Introduction

William Faulkner explored some of the themes in this book in *Absalom, Absalom!*, published in 1936. The protagonist of that magnificent novel, Thomas Sutpen, came to Mississippi in the 1830s because he was fleeing his past, his family, and the hierarchical society of Virginia, which was so old and decaying that its thick muddy rivers seemed to stand still or even flow backwards. He had a "design," which he pursued relentlessly: he wanted to be a rich planter and master at any cost. He was a brutal slaveowner, driving his slaves without mercy, and his ambitions were so consuming that he worked with his slaves in the field—something that few planters, especially in old Virginia, would have done. He also exploited his female slaves in the most intimate way, fathering at least one mulatto child with a slave woman.

Sutpen wanted to found a dynasty in Yoknapatawpha County, Mississippi, so he married a white woman, Ellen Coldfield Sutpen, and they had children together. (He already had one son from a brief marriage in the West Indies to Eulalia Bon.) Yet his driving ambition destroyed his wife and children in ways that he never seemed to comprehend. Sutpen dominated the household, which was insistently masculine in tone, lacking proper carpets and curtains, and it remained curiously unaffected by a woman's presence. He mistreated his wife, whose misery was compounded by her isolation; she scarcely saw her own kinfolk, and she became so lonely and unhappy that she eventually lost her mind.

The Sutpen family is one of Faulkner's most riveting creations, in part because the fictional Thomas and Ellen Sutpen both represent extremes of human behavior. Yet echoes of their story can be found in the experiences of planter men, who shared Thomas Sutpen's drive and dreams of wealth, and planter women, who shared Ellen Sutpen's unhappiness, even though most women did not come to her tragic end. This book is about the profoundly different ways that planter men and women experienced migration from the Southern seaboard to the frontiers of the Old Southwest in the years between 1810 and 1860. Migration was a family venture in the sense that men and women both took part in it, but they went to the frontier with competing agendas: many men tried to escape the intricate kinship networks of the seaboard, while women tried to preserve them if they could. On the Southwestern frontier, the planter family underwent "nuclearization," for want of a better word: men, women, and children found themselves alone, far from the many collateral relatives who populated the seaboard. The frontier family was not an egalitarian institution, however; in fact, migration altered sex roles and heightened the sexual inequality within the family. This book, then, is as concerned with change over geographic space as it is with change over time.[1]

It naturally enters into a rich historiography regarding the planter family, women's history, the modernization of Southern society, and the American frontier. Most scholars portray the antebellum planter family as patriarchal in its distribution of power and as nuclear in structure.[2] The seaboard family was indeed a patriarchal institution that subordinated young men to middle-aged and older men, and women to men, but it was by no means a nuclear unit; its structure was elastic, including many relatives beyond the nuclear family. Furthermore, pacts between generations of men and between the sexes ameliorated many of the family's inequities. Young men who contributed to the family's welfare would receive in exchange financial security; women were able to form interdependent relationships with many female kinfolk, which provided alternatives to the inequitable relationships that prevailed between the sexes. These pacts permitted the family to function reasonably well. Young men who migrated to the frontier withdrew from the pact between the generations, and when individual nuclear families became isolated on the frontier, the pact between men and women also broke down. The worst aspects of the patriarchal family emerged, while women were left without the sustenance of their female kinfolk.

Historians have also debated the values and outlook of women in the planter class, and they have reached varying conclusions. Some scholars argue that women's values differed from men's, even on the issue of slavery,[3] while others emphasize the influence of class rather than gender, arguing that planter women prized their elite social status, shared the values of planter men, and in general upheld the social system.[4] Until the migrations of the 1820s, 1830s, and subsequent decades, men and women agreed on the supreme importance of family connections in their lives. When the generation of migrants rejected this assumption, a gulf opened between the sexes where before there had been merely tensions and resentments. Under the stresses of migration, planter men and women became deeply divided about fundamental aspects of family life.

Many planter women, for instance, strongly disapproved of the greed that they thought motivated men who went to the frontier, and they believed that the preservation of family ties was far more important than the pursuit of riches. They often remarked on the destructiveness of migration, and their letters can make for painful reading as they describe the collapse of kinship networks that had provided them with practical assistance and emotional fulfillment in the seaboard. When they tried to make friendships with women in the new country, they were more concerned with personal qualities, unlike most men, who were primarily concerned with wealth.

Futhermore, planter women and men differed in their relationships with slaves after they arrived on the frontier. In the seaboard, many members of both sexes tried to observe the pact between the races that historians call "paternalism"—the idea that slaves were human beings and that masters and mistresses were obligated to treat slaves with a minimal amount of decency. But many planter men in the generation who moved to the Southwest rejected this view of slavery and felt little or no personal obligation to slaves.

Many planter women who went to the frontier, however, continued to practice a female version of "paternalism." They tended to perceive slaves as individual human beings more often than men did, and they commented more frequently than did men on the trials and adjustments slaves underwent as they migrated to the frontier. A few planter women even sympathized with and identified with slaves as individual human beings.

There were limits, however, to what planter women were willing to face about the nature of their society. They did not assail slavery or the patriarchal family, and only a few, at moments of great unhappiness, considered the power relationships that governed race relations or underlay the family. At a number of points in the migration process—during the family's debate over the issue, on the westward journey, and amidst the loneliness and hardship of life on the frontier—they articulated their bewilderment, anger, and despair about the dramatic changes in their lives. They usually confided in other women, however, rather than confronting the men who wrought these changes. Women did not directly challenge the social system because they had few resources, little power within the family, and almost no alternatives to a life outside the family.

Scholars have long disagreed as to whether the antebellum South was an essentially premodern or a modern society. This study shows that modern elements existed within seaboard culture even before families began migrating westward; for example, planter men already experienced mobility in a more modern fashion than did planter women. But the family still largely determined social status, and women and older men feared the drastic changes that long-distance, permanent migration might bring into their lives. As Eugene D. Genovese suggests, the South was a hybrid, a prebourgeois society tied to a modernizing capitalistic world. Planters' sons were trying to force the South to become modern when they went to the frontier—although they would not have used these terms. Migration to the Southwest was itself a form of modern mobility, a permanent relocation in a new place far from home, and young men were eager, at least at first, to break with the family and its kinship networks and form more impersonal, commercial relationships with their peers. Many of the young men I studied failed in their efforts because the necessary financial structures, principally a sound banking system, did not exist to allow migrants to fulfill their goals, and they had to return to the family and its kinship networks as the primary economic institution in the region. Their attempt and their failure both highlight in different ways the peculiarly hybrid nature of the antebellum South.[5]

The Southwest was also part of the West, and of all of the American frontiers scholars know the least about this frontier. Wilbur J. Cash envisioned the entire South as a frontier, but it is clear that most planters, and probably most whites, conceived of the region as divided into a long-settled, conservative seaboard and a raw, dynamic frontier. Planters' sons resembled other American pioneers in their optimism, their economic aspirations, and, as Frederick Jackson Turner suggested, their intense desire to escape from "the bondage of the past." Turner also realized that the westward movement had a destructive side that went hand in hand with its liberating dimensions, something that historians occasionally overlook. The settlement of the Southwest, perhaps more than that of any other antebellum frontier, confirms this double-edged aspect of the thesis—although not in ways that Turner predicted. The westward movement was exhilarating at first for the planter men who led it, but it was destructive for the planter women who went with them. Among Southwestern planters, sex roles were probably more rigid than in any other group of American settlers historians have studied so far, and the inequality within the nuclear family was probably correspondingly greater. The exploitation of enslaved people was inherent in the settlement of the Southwest as it was nowhere else, crucial, in fact, to the fulfillment of the goals of planter men. Even more than planter women, slaves were victims of the aspirations of these white men.[6]

This study is confined to families who owned at least twenty slaves—the standard definition for the planter class—and who had lived in the South since the time of the Revolution, if not longer, to ascertain that they were really "Southern." Three states, Virginia, North Carolina, and South Carolina, represent the seaboard; Maryland is excluded because slavery played a distinctive role in its political economy and Georgia because it straddled the seaboard and frontier regions. The Southwest is defined as the states and territories west of the Alabama-Georgia state line, as well as frontier areas in Tennessee, Kentucky, and Florida. The focus of this work is on men and women who made permanent homes in the Southwest, although it discusses a few individuals who stayed behind or who migrated only to return to the seaboard. The book includes some simple statistics on household structure and sex ratios, as well as the slaveholdings of migrants, their brothers, and their fathers, but the foundation of the book is the personal testimony of planter men, planter women, and slaves.[7]

1

The Ties of Nature: The Planter Family in the Seaboard

In the early 1830s, Samuel Townes of South Carolina was a restless young man. Born in 1806 in the hilly upcountry near the village of Greenville, he was the second son of Rachel Stokes Townes and Samuel Townes. His father, who was born in Virginia, moved from Virginia to South Carolina in the 1790s and became an affluent planter and slaveowner. Both of his parents' families had been in the South since the mid-eighteenth century, and they were connected by blood and marriage to a host of kinfolk in the seaboard. The household was often filled with visiting relatives, usually Samuel's aunts, uncles, or cousins from the Carolinas, as well as kin who came from as far away as Virginia. While Samuel was attending the University of Virginia in 1826, his father died, and when the young man came home he inherited three slaves. Meanwhile, he was recruited to help his male kinfolk in their business dealings. In 1829 he began studying law in the town of Abbeville, South Carolina, about forty miles from his home, with Armistead Burt. Martha Calhoun Burt, Burt's wife, was a sister of Samuel's sister-in-law Lucretia Calhoun Townes, as well as a niece of one of the South's most famous politicians, John C. Calhoun.

Samuel was part of a large, successful family, but he felt stifled in its embrace. Articulate, hard-driving, and high-strung, he was also very ambitious. He initially enjoyed his apprenticeship with Burt because "I feel and act like a man and am received and treated as such," but he began to resent the way Burt simply assumed he would join the firm without discussing the matter with him. Gradually Samuel came to distrust his kinsman, whom he later described as "cunning," and he feared his temper after Burt beat up another relative, John A. Calhoun, over a business dispute. Samuel also resented his conservative older brother Henry (the husband of Lucretia Calhoun Townes), a physician and farmer who was full of advice about the course his life should take, and he envied other successful male relatives. He dreamed of becoming a skillful lawyer, an influential journalist, and most of all a rich planter, but he believed that he had no future in South Carolina, where the soil was exhausted and a generation of older men blocked his way. In 1833 Samuel decided to leave Abbeville—which had become a "miserably dull and uninteresting" place—for Alabama. Early the next year he married Joanna Hall, a twenty-two-year-old woman from Charleston, despite his family's vocal opposition to the match, and left with her and his slaves for the Southwest.[1]

Samuel Townes was representative of his generation in several important respects. Many other planters' sons experienced family life as demanding and engulfing, and they believed that it fostered too many obligations among far too many relatives. Their parents, however, found familial assistance indispensable in running households and plantations, and they maintained emotional ties with many kinfolk. Although these men and women experienced family life in different ways, they joined together to create elaborate kinship networks that fostered an intricate web of reciprocal, interdependent, and dependent relationships. Moreover, the older generation felt a powerful attachment to their homes and their relatives' homes in the seaboard. The family was of paramount importance, even for those who wished to escape it.

The Structure of the Family

The planter family was so encompassing partly because it was not nuclear in structure. In the nuclear family, parents and children reside together in the household, and they give their deepest emotional loyalties to each other. The nuclear family is also "a state of mind," notable for the solidarity among its members and their isolation from other members of the communities in which they live. Seaboard planters and children usually lived together under one roof, and they loved each other very much, as hundreds of their letters attest. The intensity of these relationships has led most historians to conclude that the planter family was indeed nuclear in structure.[2]

But planter households also contained other relatives, and parents and children did not expect emotional fulfillment solely from relationships with each other. The family had a nuclear core of parents and children, but its core should not be confused with its social and emotional boundaries. Many individuals whom historians portray as marginal members of the larger kinship network were actually significant members of the family. Planters considered cousins, aunts, uncles, nieces, and nephews to be important relatives, and these relatives were intimately involved in each other's lives.

Many, perhaps most, planters lived in households that included individuals from beyond the nuclear core as permanent residents. The federal census returns for 232 planter households from six counties in Virginia, North Carolina, and South Carolina for the years 1810, 1820, and 1830 reveal that about one-quarter of the households (sixty-three, or 27 percent) contained families that may be defined as nuclear—those with two adults and children, whom we can assume to be parents and offspring. (See Table 1.) Slightly over one-third of the households (eighty-seven, or 38 percent) contained families that are complex in structure—those with at least three adults, only two of whom could be parents, or with some other combination of individuals that departed from the nuclear model. The remaining eighty-two households (35 percent of the total) contained families whose structure is ambiguous—the age and gender of the various household members are such that they cannot be safely classified as nuclear or complex.[3] Although the average household size for the 232 households was seven persons, some households were enormous: household size ranged from two to twenty-three persons.[4]

The short-term residence patterns of planter households also departed from the nuclear model. Most households were more congested than the census returns indicated, because relatives from beyond the nuclear core visited frequently; household membership

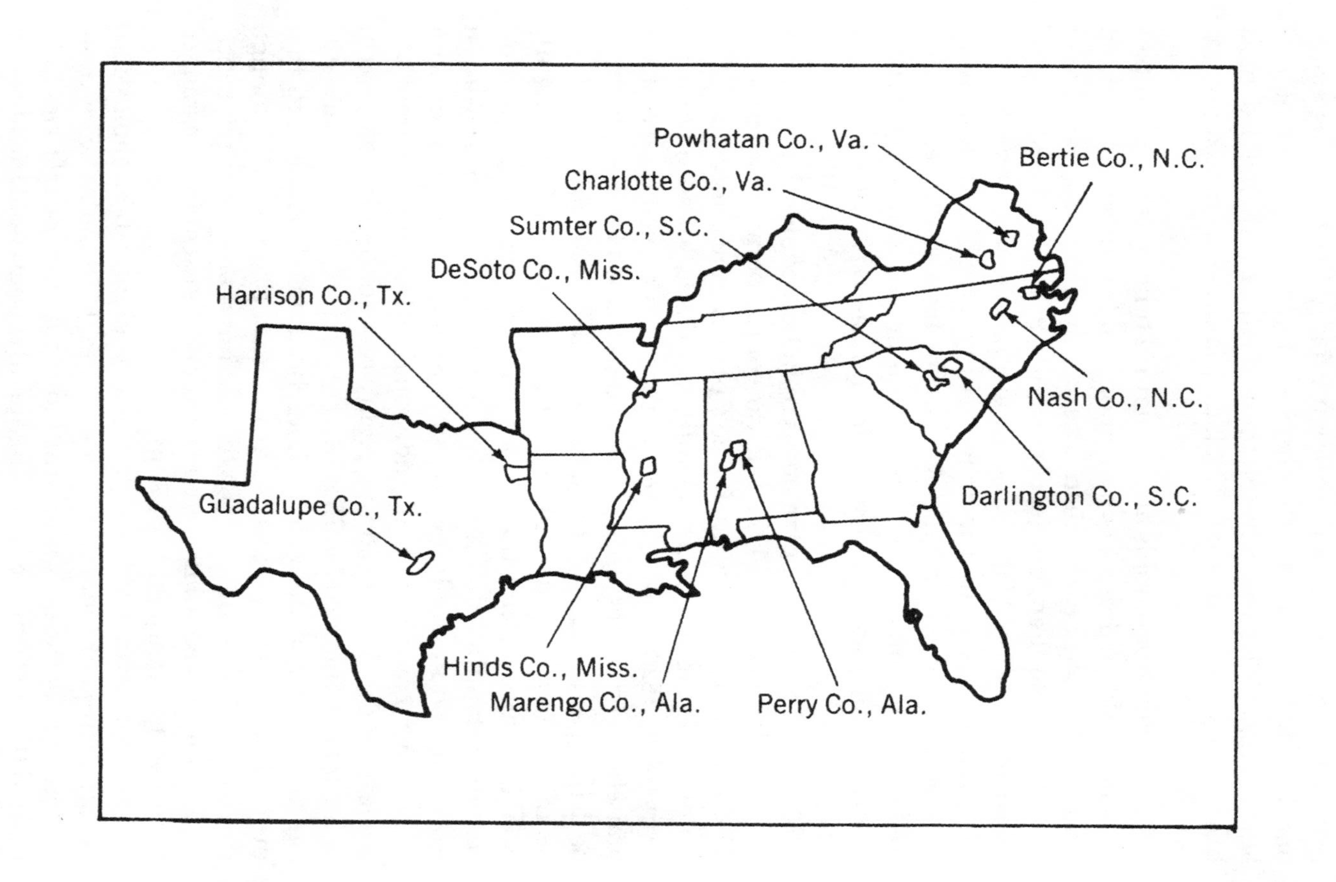
Powhatan Co., Va.
Bertie Co., N.C.
Charlotte Co., Va.
Sumter Co., S.C.
DeSoto Co., Miss.
Harrison Co., Tx.
Nash Co., N.C.
Darlington Co., S.C.
Guadalupe Co., Tx.
Hinds Co., Miss.
Marengo Co., Ala.
Perry Co., Ala.

fluctuated from one month to the next and from one year to the next. Even more important, the custom brought individuals of all ages into prolonged contact with many relatives, promoting emotional bonds with kin beyond the nuclear family. Visits were not mere formalities, but a vital part of family life. Historians have overlooked the significance of these connections between households, which were just as important in planters' lives as what happened inside of individual households.[5]

Planters made several different kinds of visits with their relatives. Relatives who lived near each other made informal, daily visits, which brought kinfolk from beyond the nuclear core into the routines of the household. Kin visited for several hours or a single day, appearing unannounced or sending a message by a neighbor, relative, or slave communicating their intentions to call. Among the Fairfax and Cary families of Virginia, aunts, cousins, and other relatives arrived without advance notice and stayed a few hours or all day before returning home. This pattern of hospitality was a virtual "law" in the family, one woman recalled, and kinfolk considered it a privilege to be exercised by anyone with the "same blood in their veins." Such an easygoing familiarity characterized relationships among many relatives from neighboring households.[6]

During ceremonial visits, kinfolk gathered to mark key events in the family's history, such as weddings, commencements, and funerals. For instance, Alicia H. Middleton reported that her sister was soon "surrounded" by "beloved relatives" who consoled her after the loss of her son. These visits drew together a range of nuclear and extranuclear relatives from throughout the seaboard, and they brought people together during times of high emotion. By their very nature, ceremonial visits confirmed a sense of shared history and identity among relatives.[7]

Kinfolk from distant counties or other seaboard states made extended visits, which had an even more important impact on the structure of the family, because the household was filled with extranuclear kin for long, uninterrupted periods of time. Unconnected to any special event, these meetings were intended primarily to cultivate affectionate, lasting bonds between kin, what one planter called "the pleasures of relationship." Merely spending time together did not always ensure this result, of course; familiarity sometimes bred contempt. Louisa Cunningham, for example, breathed a sigh of relief when a visitor left after "five long weeks" in her household. It could also breed resentment. Elizabeth Blaetterman became incensed because her husband spent so much time with his sister's family, "whose every action I fancied he preferred to mine." But planters expected that visits would result not in ill will but in stronger feelings of love and duty. They attempted to put aside rivalries, whenever possible, for the duration of the visit.[8]

Planters had the material resources—what Henry Townes once called "an abundance of room & provisions"—to take in many guests for long periods of time. Constance Cary Harrison's grandparents put kinfolk into the numerous spare bedrooms of their spacious mansion. The Lenoir family of North Carolina resided in a less elegant home, a ramshackle old house constructed in the 1780s and nicknamed "Fort Defiance" for a nearby Indian fort, but it sheltered many visiting relatives throughout the antebellum era. Some households became quite crowded, such as the home of South Carolinian Arthur Middleton, who once hosted twenty-two people (fifteen children and seven adults) for a month of "joyous intercourse."[9]

The premium during these gatherings was on conversation, the "interchange of opinion, long confidential talks, family news." Men and women renewed a sense of the family's history and passed it on to the next generation. Adam Dandridge looked forward to a visit from his sister and her family so that they could "recall together scenes of the past—and look together into the future." Planters also enjoyed picnics, concerts, and dances together. Adults, teenagers, and children also went hiking and boating, fished, and rode horses, and small girls and boys played together. Guests made visits to other kin nearby, taking tea or dining together, or traveled to local springs or resorts. On longer visits, however, the hosting family returned to its routines and incorporated relatives into its ordinary activities. Children studied together, women assisted each other in housework, and men accompanied each other on their daily business.[10]

Many of these visits succeeded in creating durable emotional bonds among a variety of relatives from different households. Mira Lenoir found herself "very much attached" to a cousin and her new husband after they spent a month in her home, and Mary Ann

Taylor related that her aunt "endeared herself to me tenfold by a thousand nameless acts of kindness during my residence with her last summer." One planter's son was devoted to his aunt, who "from my infancy, . . . treated me as a son," while another told his aunt that he would never forget her "motherly kindness while I lived with you."[11]

Although both men and women loved their relatives, the sexes nonetheless experienced the emotional aspects of visits in different ways. The lives of women revolved around the family, and if one group of kin visited more often than any other, it was probably the plantation mistress's female relatives. During visits they sought the companionship and intimate conversation that were not always evident in their relationships with men. A young North Carolinian enthused that her cousin's letter produced "a kind of 'heaven upon earth'" and added that the prospect of seeing her was "the hope that cheers me onward." Another woman planned a visit through the seaboard for "as far as the relations extend." Many women intensely enjoyed the time they spent together. Addie and Carrie Dogan savored the "many joyous hours" they spent with their cousin during one summer-long visit; years later, one of them remembered her seasonal visits as "blissful."[12]

The customs of visiting in fact highlighted women's geographic immobility. While planter men traveled at will—walking, riding on horseback, or taking a carriage—planter women always had to travel with male escorts. Men guided women through the discomforts of travel and protected them from possible danger; a man's presence also signaled that his companion was a respectable woman, part of a family, and inaccessible to other men. Martha Lenoir Pickens was accompanied by male relatives or family friends as she journeyed within North Carolina in the 1810s, and the women of the Calhoun family always traveled with male escorts in the 1820s and 1830s. These constraints on women's mobility lasted throughout the antebellum era, and they obtained for distances long and short. In 1846 a Virginian plantation mistress was unable to visit a guest on a nearby plantation because she could not persuade her husband or her brother to drive her there. A woman's mobility, then, was in every way dependent on the consent of a man—his permission to allow her to travel and his willingness to accompany her to her destination.[13]

Some women felt frustrated by their immobility, such as the matron who became "exceedingly angry" when her son did not bring her daughter-in-law and grandchildren to visit her. Others no doubt envied the easy mobility of men, like the North Carolinian who wrote wistfully to her cousin, "If I was a young gentleman I would go to see you." Women occasionally defied their male relatives, as did Mrs. C. W. Downing, who struck out on her own to visit relatives near her Virginia home after she quarreled with her husband, but very few women were willing to flout the wills of their husbands, fathers, brothers, or sons and go forth from the plantation alone. Most tried instead to accept the situation and visit as often as their male kin deemed it convenient.[14]

Fortunately, many planter men valued the high sociability of family life and facilitated visits much of the time. John B. Dabney enjoyed the "intimate and endearing intercourse" among his kinfolk in early nineteenth-century Virginia, when his grandfather's brother's home was the "common focus" of family visits. But men's lives were never so completely involved with the family; they had other social relationships in the public and professional realms. The custom of visiting flourished because men allowed it to flourish, but it was always more central to women's lives than it was to the lives of men.[15]

These families were other than nuclear in structure, but they nonetheless fostered the emotional bonds usually associated with the nuclear family. Planter men and women resided intermittently with their kinfolk throughout their lives, and they loved and sometimes hated many relatives beyond the nuclear core of the family. A complex family structure, in other words, coexisted with deep emotional bonds.

Kinship

Underlying this kind of family life were certain demographic, geographic, and social conditions. Mortality rates declined in the early nineteenth-century South from the treacherous levels of the colonial era, but birth rates remained high. More parents lived to see their children reach adulthood, and more children survived to raise families of their own; more collateral relatives also survived.

These changes allowed planters to participate in large kinship networks, an experience that had been less common in the seventeenth and eighteenth centuries.[16]

Furthermore, seaboard relatives frequently settled near each other, so that individual families lived surrounded by kinfolk. The Dogan family was related to at least fifty individuals who lived in their native Union County and nearby counties in the South Carolina Piedmont. The Gordons and Hacketts had kinfolk in twenty-one households located along the Yadkin River in three counties in western North Carolina. Leah McFadden, born in South Carolina in 1771, had 130 descendants by the time she was eighty-three years old, and almost 100 of them resided in their native Sumter County.[17]

Planters helped preserve kinship networks by collecting genealogical information about their ancestors and their living relatives. When William Henry Holcombe compiled his genealogy, he consulted his female kin, whose knowledge covered more than one hundred years of the family's history. George Dromgoole sent a genealogy to his sister for her approval, and she supplied details from papers pertaining to their grandfather's will. Women also actively solicited this information. One man told his cousin that his wife had "another question for you to answer"—she wanted to know the name of the child of another cousin.[18]

Finally, planters believed that kinship ties were a part of nature itself, ordained by God and therefore sacred. M. E. Patterson declared that she would always cherish her cousins, who were "allied to me by the ties of nature" as well as the bonds that existed between "our forefathers." Another couple took in an orphaned nephew, who was "very nearly allied to us by the ties of nature," because it was their "sacred duty." William H. Holcombe called his father's love for his family and his native Virginia "the most sacred ties of his being."[19]

These kinship networks operated on the principle of reciprocity: the assumption that relatives should help each other manage plantations and households and support each other through the vicissitudes of life. Both sexes assumed that kinship was based on the ties of nature, and both derived strength, support, and plain enjoyment from these relationships. But men and women acted upon the principle of reciprocity in dissimilar ways, according to the resources they had access to and the degree of autonomy they enjoyed.[20]

Men saw the family as an economic unit, as well as a group of individuals bound together by blood, marriage, and emotion. Other sources of capital and credit were scarce, and the banking system of the seaboard was not always adequate to meet the needs of planters who lived on credit between harvests, so men turned to their male kin for economic assistance. These relatives sometimes charged interest on their loans (typically 5 or 6 percent, the prevailing rate in the seaboard in the 1830s), but relatives were usually generous creditors, and most men preferred them to bankers, friends, or acquaintances.[21]

Planters made gifts and loans to an array of relatives, including cousins, in-laws, uncles, nephews, grandfathers, and grandsons, as well as fathers, sons, and brothers. They exchanged slaves, land, and money and provided collateral or "went security" for each other's loans and investments. They also did more ordinary favors for each other, such as lending a carriage for a journey. Some of these favors involved face-to-face negotiations, but many could be transacted through the mail between men who lived in different counties or different states; planter men's correspondence is filled with requests for such favors and thanks for favors received.[22]

Men became involved in these relationships not only because it was good business, but also because they felt obliged to help their kinfolk. Charles Cocke felt "moral as well as legal obligations" to pay a debt and help out a cousin who needed money. Some planters could be very generous. An affluent old Virginian was known for his "entire willingness" to assist his grandchildren "in any reasonable enterprise." Men expected their kinfolk to aid them if they got into trouble. When Ellis Malone feared that his brother was bankrupting their father's estate, he turned to his uncle William Lea for help.[23]

Planters took reciprocity seriously, so seriously that they believed obligations should be handed down through the generations. Peter MacIntyre, a North Carolinian who exchanged many favors with male kinfolk, proclaimed that his motto was "An old friend and your father's friend forsake not." Hard feelings resulted when men failed to honor these relationships across generations.

C. W. Downing became indignant when relatives refused to help him, declaring that "many *should* have done it on account of favors shown them" by his father. But most men tried to preserve these systems of assistance, despite the conflicts that arose within them. They accepted interdependent relationships as natural, right, and necessary to accomplish the many tasks of running plantations and businesses.[24]

Women also developed reciprocal, interdependent relationships with their female kinfolk, exchanging gifts and favors that were smaller, but no less important, than the propery and favors that men exchanged. They gave each other food, clothing, and books, and they relied on each other for help in running households, raising children, and caring for the sick. The first cousins of Mary Custis Lee (Mrs. Robert E. Lee) helped her iron clothes and chaperon children on outings. Two Virginians spent a fortnight caring for their ailing kinswoman, providing both "considerable service" and "good company." Relatives also gathered when a woman was about to give birth, as when Elizabeth Barbour Ambler's mother and sister joined her before she went into labor.[25]

Women sought less specific aims in their relationships with female kinfolk—emotional well-being and "good company." They wanted each other's personal presence so they could share their feelings with each other. For example, a matron was expected to visit her grieving cousin and "stay all night and try and talk her into cheerfulness." If women could not be together, they poured their thoughts into their letters. Lucretia Townes loved the missives of her sister-in-law Eliza Townes Blassingame because they were "almost equal to verbal chit chat." Women's relationships with each other did not always run smoothly, of course; the letters between Julia Pickens and her aunt and guardian Mira Lenoir simmered with mutual distrust. Lenoir demanded that Pickens write frequently and tried to force expressions of affection out of the young woman—affection that Pickens obviously did not feel. The two women nonetheless carried on a strained correspondence. Women, like men, tried to suppress conflict and preserve relationships whenever possible.[26]

Kinship networks also provided, to differing degrees, the foundation for the social networks of men and women. Relatives were the most important constituents of the social networks of both sexes, but men's social networks contained many individuals beyond family. John J. Ambler, Jr., of Amherst County, Virginia, was the proud son of a rich planter, and he knew relatives throughout the state in the 1820s and early 1830s. But in his daily life he also met former college classmates, planters, professionals, friends, and strangers, recording transient meetings with hundreds of individuals in many environments. Women's social networks more often coincided with their kinship networks, and as a result their social networks were smaller and more stable. They maintained friendships with their classmates from school, their neighbors, and women they met at church, but the "central places" of their lives were the residences of kinfolk. Elizabeth Barbour Ambler, John's mild, quiet wife, spent most of her time with about two dozen kinfolk who lived near her home in Amherst County and her parents' residence in Orange County.[27]

The geographic scale of men's social networks was also larger than women's. John Ambler traveled throughout the Southern seaboard and the Northeast on business, and many other planter men were well-traveled individuals; their social networks covered hundreds and even thousands of miles. Women, however, rarely traveled beyond the counties where their kin resided. In the Pickens and Lenoir families women circulated in the 1830s and 1840s between their homes in four counties in central North Carolina and those of several cousins about one hundred miles away in upcountry South Carolina. (Here there were "enough ladies . . . to keep you visiting all the time," according to one woman.) A few women maintained social networks over great geographical distances. Mary Cox Chesnut visited relatives in Philadelphia annually after she married South Carolina planter James Chesnut, Sr., but she was a native of the North, not a typical plantation mistress. Men, not women, experienced mobility as easy, linear movement across geographic space. Women experienced mobility as circular rather than linear, and they moved in familiar paths in a smaller world of familiar faces.[28]

Child-Rearing and Sex Roles

In most families mothers and fathers were the primary caretakers of children, but they expected relatives beyond the nuclear core to share in the responsibility. Many adults from the greater kinship network accepted and even sought these duties, so that a number of individuals helped raise the next generation. One planter expressed the views of many parents when he said that he wanted his son to associate with "the same sort of people with whom I have been raised."[29]

Neighboring kinfolk frequently provided advice, counsel, and guidance for planter children. James Ewell Brown ("J. E. B.") Stuart regarded his uncle as his best friend, a loyal, sympathetic mentor. Mary Boykin Chesnut's grandmother was a second mother to her, and young Mary was "her Shadow," "her pet," following her around the house as she did her chores. William H. Fitzhugh of Virginia enjoyed an "intimate acquaintance" with his kinsman Robert E. Lee in the 1810s and 1820s and helped him gain admission to West Point Military Academy. These individuals were a vivid presence in the lives of young people.[30]

Planter children also had sustained contact with their relatives through the practice of child exchange, as siblings, in-laws, aunts, uncles, nieces, nephews, and cousins throughout the seaboard sent their offspring to each other's households for lengthy visits. Adults took in children for a variety of practical reasons. One planter sheltered his ailing nephew and promised to "do all we can for his comfort," while others took in their young relatives every summer to safeguard their health. Parents also wanted their children to enjoy opportunities during visits they would not ordinarily encounter. A North Carolinian "took a great fancy" to his cousin's son and asked that the boy be allowed to live in his own household. The planter promised that the boy would have the best education available and other privileges that he could provide.[31]

The most important functions of these visits were broadly social, however, rather than practical. As a means of socialization, child exchange resembled the practice of apprenticeship in early modern cultures. Planter children did not work as servants in other households, of course, but they did absorb a sense of identity, the knowledge that they were part of a kinship network. Stephen D. Miller of South Carolina was raised "principally among my Mother's relations," and he grew up believing that they were "most worthy of affection and the cleverest people in the world." Other children learned to love their relatives, such as the sisters who told their cousin at the conclusion of a long visit that they "did not know how fondly we loved you, until you had gone."[32]

Many parents also believed that visits with kinfolk helped shape the characters of their children. One man taught his brother's grandchildren "sound moral and religious principles" and "wise maxims" during their many visits to his home. Parents tried to teach girls and boys to refrain from excessive behavior, according to one North Carolinian, to practice moderation in all of their affairs. John C. Calhoun had grown up in the "habit [of] intimacy" with his own cousins in South Carolina, but his political career kept his offspring in Washington, D.C., during much of their childhoods. He confided to one of his cousins that his inability to raise his children among their relatives was "quite a misfortune." He declared, "I . . . love the tie of relationship and believe that those who are under its influence are usually much more disposed to a virtuous life."[33]

As boys and girls approached adulthood, they learned to function as members of these large kinship networks. They began to give to as well as receive from a body of relatives, but the sexes absorbed different lessons appropriate to their roles in the family. Once again, many members of the kinship network helped young people make the transition to adulthood, and they stood ready to punish young people who violated precepts of behavior.[34]

Since boyhood, planters' sons had been acquiring traditional masculine skills, such as hunting, fishing, and riding, from their fathers and other male relatives. Young men also learned to defer to their older kinfolk and to be honest in dealings with them. It was important, as one man warned his son, to retain their "good wishes," because kinfolk withdrew their support when young men repeatedly misbehaved. Members of the Dabney family, for instance, ostracized one kinsman who gambled away his fortune and went bankrupt. Relatives did not always succeed in enforcing correct behavior, but their disapproval was something to be reckoned with.[35]

The young adulthoods of planters' sons typically followed a certain sequence. They graduated from academies in their middle to late teens, attended college for several years, and then returned home, where they studied law or medicine or simply lived with their parents or with a kinsman. Many learned to drink, if they had not already done so, and others began or continued sexual relationships with slave women. One graduate listed his pastimes as follows: "hunting, playing, working, wagoning, whoreing, courting, and the Devil knows what all."[36]

There followed an interval of five, six, or even ten years of semidependency, in which young men lived with their relatives while they prepared for occupations, courted, and married. These residence patterns are visible in the census returns for 1830. One-half (51) of the 103 households surveyed in Table 1 contained men between the ages of twenty and thirty who were living at home with older adults presumed to be their parents.[37] The passage to adulthood was slow and haphazard, as young men accumulated property, status, and dependents of their own. Along the way they passed certain landmarks, such as reaching twenty-one years of age, passing the bar, and marrying, but there was no clear turning point after which a boy could be said to have become a man. A Virginian approaching his twenty-first birthday felt that he was beginning to "lose this boyish feeling" but was still "not ready to be a man." Many planter sons did not establish households of their own until their late twenties or early thirties, when their fathers placed them on land nearby or sent them into the professions. A young man was only "somewhat" in control of his fate, as one plantation mistress observed.[38]

In this state of semidependency, young men began to participate in the systems of economic exchange that flourished among their kinfolk. They ran errands for their older relatives, helped supervise plantations, and purchased slaves and livestock. They assisted their fathers and brothers as well as uncles, in-laws, and cousins from both sides of their families. For example, Richard R. Hackett's cousin hired some of his slaves, and when the cousin traveled from home, Hackett in turn supervised his plantation. Young men were supposed to be learning a central aspect of the traditional male sex role, namely that men had duties to the family at large that were more important than their individual ambitions.[39]

Older male relatives, especially fathers, continued to be powerful figures in young men's lives. Fathers kept a close watch on their sons' courtships, and some did not hesitate to intervene in romances they thought inappropriate. Robert Cunningham of South Carolina stopped his son John from marrying the woman he loved because she was from a poor family. Some fathers chose occupations for their sons, who accepted whatever they decreed. A South Carolinian ordered his son to "go to the Dunlap place to farm with five hands [i.e., slaves] and the half of what you make will be yours." As one young man noted, planters' sons were "bound to consult and comply with" their fathers when making decisions.[40]

Some young men shared the feelings of the South Carolinian who denounced the "tyranny" of parents who interfered with courtships, but they had to accept this despotism if they wanted the security older men could provide. This was how the pact between generations of men worked: young men who bided their time, respected their elders, and served the family would be taken care of, as Kimbrough DuBose discovered. He felt trapped by obligations to his kinfolk, but when debts brought him to the edge of ruin, his relatives saved him, buying his slaves at auction and giving him a plantation to live on free of charge while he recouped financially.[41]

The upbringing of planter daughters differed in most respects from that of their brothers. Other scholars note the belletristic education that parents gave to their daughters, the careful attention to their morals and manners, and the general preference that many parents had for sons. Daughters received gifts of property from their parents, typically several slaves when they married, but they usually did not have access to the substantial amounts of property their brothers received, and many were uninformed about the size and disposition of the family's wealth. They lost the right to control whatever property they owned when they married, since Virginia, North Carolina, and South Carolina denied property rights to married women in the antebellum era.[42]

Furthermore, parents taught girls to nurture and maintain familial relationships, responsibilities that increased as girls grew older. As teenagers they expanded their circle of correspondents and began inviting kinfolk for visits. They also learned the skills of household management from mothers and other female kinfolk, so that by the time most girls reached their early twenties, they had learned enough to run establishments on their own. When Joseph and Sarah Dogan traveled from home in Unionville, South Carolina, they left their daughters Addie and Caroline, twins who were twenty-four years old, to act as "housekeeper, hostess, mistress, nurse etc." until they returned a few days later.[43]

As young women learned how to run households, they too learned to defer to older relatives, who scrutinized their behavior during visits. Anna Calhoun Clemson, daughter of John C. Calhoun, observed that her young cousin's manners had "improved" and that she was "very amiable." Women, like men, shunned those who did not conform to proper standards of behavior. Members of the Calhoun and Simkins clans avoided a woman who was crude and boorish. Men too passed judgment on the conduct of their female kin. A North Carolinian remarked that his kinsman's daughters "cannot set the river on fire yet"—in other words, they had more to learn about proper female conduct.[44]

Women in their late teens and early twenties were primarily concerned with marriage, of course, one of the chief events of their lives. Young women could choose among suitors and influence their own destinies to some extent. Beginning in the mid-eighteenth century, some young women chose spouses according to their personal preferences rather than the wishes of their parents, but this does not mean that parents had no authority in the choices their children made. Jane Harris's father refused to permit her to marry Joseph Woodruff, a wealthy army officer she met in South Carolina in the 1810s, but relented several years later when she admitted that she still loved the man.[45]

Newly married couples typically lived for several months or even several years with relatives and then settled in a home of the husband's choosing, a separation that often disrupted close associations between a bride and her female relatives and friends. South Carolinian Mary Ann T. Harwell sent both congratulations and "condolences" to a friend who was engaged because she knew how hard it was to leave home. The youth of many planter brides must have made this separation even more difficult, since the average age at first marriage was twenty and some brides were teenagers. Yet they accepted these changes as part of married life, because it was a woman's duty to follow her husband.[46]

Unlike their male kinfolk, young women in their teens and twenties did not pass through a phase of semidependency. They did not travel at will, transact business, or engage in anything like the multitude of activities in which their brothers engaged.

Women were truly dependent, living under the control of their parents with no prospect of acquiring the autonomy that awaited their brothers when they established households of their own. Their status increased after they married, bore children, and ran their own households, but their husbands made the key decisions in the family, whether it was the location of a home or the length of a visit. Even mature, middle-aged women did not obtain the autonomy enjoyed by adult men. Although women developed interdependent relations with other women, great differentials of power characterized relations between the sexes. Women as a gender were dependent on men.[47]

Just as a pact prevailed between generations of men, however, certain tacit agreements existed between the sexes. Most men recognized how important kinship ties were to women, and most men facilitated travel between households much of the time to help these relationships flourish. Women had no choice but to accept their dependence on men, but they relished interdependent relations with their female relatives. These kinship bonds provided women with valuable practical, social, and emotional resources; they compensated somewhat for women's powerlessness within the family and created alternatives to the unequal relationships between the sexes.

Slavery, Geographic Place, and Social Class

Finally, another pact existed between planters and slaves. The "paternalism" discussed by Eugene D. Genovese evolved in the seaboard; here masters first described themselves as symbolic fathers who cared for supposedly childlike slaves and portrayed slave labor as a form of compensation for their guidance—a manifestly self-serving, racist idea that attempted to obscure slavery's evils. But, as Genovese suggests, the paternalistic model predicated that slaveowners ought to treat slaves with some amount of decency, and it could prevent some slaveowners from acting on their worst instincts. It also introduced the idea of mutual obligation between whites and slaves, which "implicitly recognized the slaves' humanity."[48] Planter men already abided by pacts between generations of planter men and between planter men and women; these pacts were built upon the ties of "nature," not upon brute force, but the point is that planter men were already familiar with the concept of inequitable social relationships that contained an element of reciprocity.

Seaboard writers such as Virginian Thomas R. Dew used the family analogy in the 1820s and 1830s when they defended slavery from attacks by abolitionists. Dew and others argued that masters, like parents, would strive not to mistreat those who were dependent on them; these writers believed that slaves, like children, needed guidance. Seaboard masters seem to have been familiar with the analogy, and the phrase "white and black family" and variations of it appear in their correspondence; Henry Townes once told his mother, "My family white & black are in excellent health." Some planter men tried to fulfill its precepts, although they were still capable of mistreating slaves and certainly did not relinquish their racist assumptions. John Tayloe III stated in his will that he disliked dividing slave family members, and inventories of his slaveholdings show that he usually did not separate mothers from children. Henry Townes also tried to follow the paternalistic model. He realized that slaves were human beings, and he did not like to see their families separated by sale if it could be avoided. In 1832 he told his brother George not to sell a slave woman to a white man who lived far from her "*family & relations*," and two years later he said that the "feelings" of a male slave "should be consulted to some degree" before he was sold.[49]

Planter men and women nonetheless related to slaves in different ways, and those differences grew out of sex roles. Planter men, not planter women, held the ultimate authority on the plantation and exercised the ultimate authority over slaves. Planter men, not planter women, typically forced sexual relations upon slaves, and they routinely bought and sold slaves as most planter women did not.[50]

In daily life, however, planter women probably had more contact (nonsexual contact, that is) with slaves because sex roles dictated that they were in charge of their physical well-being. As other scholars have amply documented, planter women fed slaves, made clothing for them, and nursed them when they were ill. Mary Boykin Chesnut's grandmother distributed medicine, shoes, and clothing to her slaves "for hours" at a time. Planter women were also responsible for maintaining human relationships in the household, and it was most often planter women rather than men who reported news about the family's slaves or inquired about slaves owned by white relatives. Their influence did not extend very far into the slave quarters—a world that whites could know partially at best—but they sometimes mediated disputes between masters and slaves. And planter women more often than men occasionally sympathized with slaves' working conditions. When Caroline Jones Patterson saw two hundred slave men working in a mine near Morganton, North Carolina, laboring up to their knees in mud, she wrote, "I could but pity the poor slaves."[51]

This is not to overlook the fact that planter women usually shared the racism of planter men and that both sexes oppressed slaves as members of the slaveowning class. Nor were all plantation mistresses kind-hearted figures. Martha Burt, wife of Samuel Townes's kinsman Armistead Burt, was "always scolding and whipping" her house slaves, even the adults, whenever she felt angry. (Samuel Townes clearly approved of her outbursts.) Some slaves disliked, feared, or hated their mistresses as much as or more than their masters. Yet many planter women practiced paternalism in a distinctly personal fashion, and the model generally offered something to slaves despite its fundamental inequity. Both planter men and women agreed, however, that whites had some obligations to their slaves because slaves were human beings, even though the vast majority of whites believed that slaves were not their equals.[52]

Planter men and women also shared certain attitudes toward geographic place, developing an enduring love for the land they lived on year after year. Virginian James P. Cocke wanted his place to remain in the family "to the end of time," adding that it would "grieve" him if it ever fell into the hands of strangers. They cherished their own homes as well as the homes of their kin. A North Carolinian thought that her cousin's home, the scene of many reunions of the Pickens and related families, was the "loveliest place on earth." Planter men and women loved places that had long associations with their families, and the longer the association, the deeper the mark on the human heart. North Carolinian

Adelaide Stokes spent a New Year's holiday at the home of a relative "filled with joy and gladness" beneath the roof of "that time-worn mansion."[53]

Men and women knew the seaboard landscape in much the same way that they knew many of their relatives who inhabited it: they had an intimate knowledge of it and an emotional commitment to it. Catherine D. Edmonston lived on a North Carolina plantation that had belonged to her family since the early eighteenth century. She and her husband often walked the grounds together, and one day, as she surveyed her garden, she exclaimed, "Everything is a source of pleasure." A married couple looked forward to visiting kinfolk in western North Carolina where their parents grew up; there, according to their son, they could "tread the ground so sacred to the memory of those loved and lost ones." Another woman recalled the majestic pines and cedars that surrounded her grandfather William Lenoir's house, Fort Defiance, her "ancestral home." He planted them himself in the late eighteenth century, and they were still "magnificent" decades later, a symbol of her family's continuity. The powerful love of place many southern whites felt is a cliché among historians, but it is worth noting how attached planter men and women were to their seaboard homes.[54]

Furthermore, for both men and women, kinship ties seem to have reduced the significance of class differences among whites.[55] The factors of blood, marriage, emotion, and reciprocity that bound people to each other had always coexisted with potentially disruptive differences in wealth, and the family itself could breed conflicts that were more long-lasting and injurious than conflicts based on class distinctions. The interaction among all of these factors was a complicated one, and the balance shifted as individuals confirmed, neglected, or ended relationships all over the seaboard. Families could be drawn closer together by economic woes. One planter opened his door to relatives who "under the pressure of necessity, or some disastrous combination of events" needed a home. Others sympathized with the financial difficulties of their relatives. The Dogan sisters welcomed a visit from a cousin even though her father was at the edge of bankruptcy; they pitied her when he died, leaving the family in "very moderate circumstances," and continued to correspond with her after she moved to Charleston to run a boardinghouse.[56]

Planters were not blind to distinctions in wealth, however, and their arrogance could strain family ties, particularly ties between men. Robert Hunter offended a less affluent cousin when he refused to let a slave woman nurse one of the man's children. Relationships could also be undermined by the jealousy that success could ignite. Francis W. Pickens, a rich planter from South Carolina, made loans, supervised harvests, sold crops, and drew up wills for the family of John C. Calhoun, his first cousin once removed, in the 1830s and most of the 1840s. In these years Pickens owned between two hundred and three hundred slaves, while Calhoun owned between thirty and eighty slaves. The Calhoun sons, who never quite succeeded at anything, began to resent their cousin's wealth, and the final straw came in the late 1840s when John Calhoun broke with Pickens over a political dispute. By the time Senator Calhoun died in 1850, all friendly relations between the two families had ceased.[57]

But the unifying bonds of family typically proved to be stronger than the corrosive effects of class differences. In the seaboard family, the balance usually tipped back toward the family and its "ties of nature" rather than toward considerations based on wealth. Most planters would have seconded John Cooke, who reassured his niece that she need not be ashamed because she had to teach school for a living. She was part of one of the "oldest and most respectable families in Virginia," one known for its "unblemished integrity and virtue," and people would respect members of the family even if they became "overseers and school mistresses." Most planters would have agreed with John B. Dabney, scion of a well-connected, wealthy planter family, who believed in qualities of "blood," in the "innate and hereditary propensities" which ran in families. The "poor man in his cottage" as well as the "nabob in his palace" had the right to be proud of his ancestry. What counted, he said, was virtue and character.[58]

By the 1820s and 1830s, then, residents in the long-settled seaboard had generated a body of values and customs about family life. Planter men and women, for all of their distinct uses of kinship, accepted without question that the family was supremely important, and they tried to teach their children the same. Daughters more often than sons, however, absorbed this view of the family; because young women were largely excluded from the accumulation of property, they had no pressing need to question tradition. When the seaboard economy faltered in the 1820s and 1830s, and it seemed that the planter family could no longer guarantee security for its younger male members, a breach opened between generations of men.

2

In Search of Manly Independence: The Migration Decision

Young men who faced the diminishing economic opportunities of the 1820s and 1830s formulated a new set of social values, embracing individualism, competition, and risk-taking. They also formulated a new definition of the male sex role, which emphasized the fulfillment of a man's personal goals at the expense of his obligations to other human beings. Planters' sons wanted to be independent of the family rather than submerged in it, and they thought the traditional pact between generations of men had little to offer them in the seaboard. They believed that the best way to ensure their independence was to leave for the Southwest.

Other members of the planter family had different perspectives on migration, because it threatened the pacts between generations of men and between the sexes. Most fathers and older male kinfolk saw it as a challenge to their own authority, an abandonment of the family, and a violation of traditional concepts of masculinity, as well as a foolish financial risk. Women, with a few exceptions, opposed migration because they thought it would undermine or even destroy the family and take their loved ones to a distant, dangerous place. Migration also threatened the underpinnings of the paternalistic race relations that prevailed in the seaboard, because of the even greater danger it posed to slave families. Slaves feared moving to the frontier because it would split up their families and because of the severe living and working conditions on the frontier.[1]

The Sons

Most young men who considered migration were in their twenties, and most wanted to be planters. Some aspired to secondary careers in the law, medicine, or journalism, but, as Samuel Townes once remarked, the money was in planting; it was also far and away the most prestigious occupation in the South.[2] They believed that soil exhaustion prevented them from becoming successful planters in their native states and, because troubles in plantation agriculture affected other sectors of the economy, that there was no future there regardless of which occupation they chose to pursue. In the words of one young man, a life in the seaboard promised nothing but "mediocrity." When young men looked around them, they did not see a countryside rendered beautiful by long association with their families. Instead they saw an old, decaying landscape, what John J. Ambler, Jr., of Virginia called "one wide waste of desolation." They felt little, if any, emotional attachment to the seaboard.[3]

Planter's sons probably exaggerated the extent of soil exhaustion, but the fields of Virginia and the Carolinas had been cultivated for over a hundred and fifty years, and the wasteful agricultural methods of several generations of farmers had taken their toll. Meanwhile, the demand for cotton on the international market increased dramatically in the early nineteenth century. The Southwestern soil was fresh and fertile, and it was better suited to raising cotton than many areas of the seaboard, which were actually more hospitable to tobacco and grains. The rivers within the Southwest were also more navigable, which reduced transportation costs, and the steamboat and later the railroad drew isolated areas into the market. It seemed as if slavery would be more lucrative in the Southwest.[4]

Planters' sons created a highly positive image of the Southwest, constructed from their reading in periodicals, newspapers, and emigrant guides, as well as the accounts of relatives and friends who had already settled there. John J. Ambler, Jr., reported that one of his neighbors left Virginia in 1832 too poor to "pay his postage" but four years later was the proud owner of seven and a half miles of rich Mississippi land. The richness of the soil stimulated prospective planters, but other features of frontier life were also attractive: the numerous lawsuits over land claims and other disputes excited lawyers, and the region's reputation for fevers and general illness drew many doctors, as their half-embarrassed inquiries made clear. All prospective migrants, however, viewed the Southwest as a place of economic opportunity, a splendid "land of promise."[5]

Young men who remained in the seaboard could expect that their wealth-holdings would grow gradually over their lifetimes, but the increase in any given year would be small. The prospect of a slow, steady accumulation of wealth was intolerable to young men who were hungry for "independence," especially "manly independence," terms which recur in their correspondence. The independence they sought was partly economic; they saw migration as a shortcut to wealth and a way to bypass the interval of semidependency that awaited them in their twenties. Edmund Jones did not want to "creep and crawl in North Carolina like a poor sloth" when he could get rich on his own in the Southwest, and Joseph Dukes did not want to spend ten to fifteen years practicing law in South Carolina before he became a success.[6]

Independence also had a psychological or behavioral component, which was even more important to this generation of men. The concept had special resonance in a society in which there were so many dependent people, most obviously slaves. The idea of personal independence had roots in the republicanism of the seventeenth and eighteenth centuries and had always been linked to the autonomy provided by property ownership. In the early nineteenth century, Northern white men continued to uphold this traditional definition of independence, but in the South the presence of slavery had subtly recast its meaning. There white men came to believe that slavery did not threaten their independence, but instead guaranteed it. Slaves as a form of property confirmed the independence of every slaveowner, but the very existence of slavery also highlighted distinctions within the household between white men and those who were dependent upon them.[7]

Planters' sons wished to break free of dependency because it seemed uncomfortably similar to the condition of white women, white children, and slaves. One planter described the plight of his widowed sister-in-law: "go where she may she must in some degree be dependent." William M. Ambler, a son of John J. Ambler, Sr., bristled that his parents "will in spite of all my endeavors treat me as a boy." Slaves, even more than white women or children, provided daily examples of what it meant to remain dependent, and some young men felt so exploited that they compared themselves to slaves. South Carolinian Louis T. Wigfall, a mercurial young law student, felt outraged because he had to give up everything in pursuit of his career, and he complained bitterly that he was "hard driven like a slave." Of course, no planter's son ever worked as hard as a slave, but the analogy reflects the fear of dependency that had taken root among this generation of men.[8]

Independence also meant freedom from the family and its elaborate kinship networks, specifically the domination of fathers and other male kin. A Virginian objected that young men in his state were "kept back" by the "prejudices" of old men, in contrast to the Southwest, where they could prosper. Another celebrated the prospect of striking out on his own and escaping his "bossing brothers." Some young men felt such intense resentments that they wanted to put as much geographical distance as they could between themselves and their families. One North Carolinian reported his kinsman's departure for the Southwest "on the ground of paternal indignity" and added, "Hate could not reach half the distance."[9]

Planters' sons also criticized the interdependent relationships between men in the family, which they saw as destructive and oppressive. Louis Wigfall declared, "If I can't support myself I ought to starve." In the Southwest he would establish a reputation for "integrity and independence," and he would "flatter no man for favors and fawn on no one for his influence." Another South Carolinian believed that too many individuals clung to their families for support and thought this behavior was "subversive of the best and most useful qualities of man." He wanted to detach himself from his family and make his own choices, free from the interference of others: "Independence in action and thinking for one's self are too often lost by this mistaken mode of doing things."[10]

Young men recognized that the fulfillment of their ambitions would conflict with the needs of their families, but they believed that ambition came first. When Samuel Townes learned that his sister Eliza Townes Blassingame and brother-in-law William Blassingame were leaving for Alabama, he calmly remarked, "This world is essentially a business world," and he believed that individuals had to put aside their feelings to pursue wealth and security. A Virginian mused that he wanted to "be of some service" to his relatives yet "still do what is best for myself." He concluded that "we all act on the selfish principle, and I suppose I must be excused for doing as other folks do." Henry E. Blair described more poignantly the choice between family and ambition. He sympathized with the "burning sorrow and anguish" his migrating cousin felt, especially the "struggle of his generous and noble heart, between the demands of his aspiring ambition, and the love of home and friends." But he thought his cousin did well to go to the Southwest, where success awaited him.[11]

Planters' sons believed that they could demonstrate their masculinity more effectively in the Southwest. They would be tested to the utmost on the frontier, a place of great struggle, and the struggle would "transform the boy into the man." Henry Townes, the conservative older brother of Samuel Townes, called it a "theater" where a man had to rely on himself alone. Many young men welcomed the chance to act in such a theater. One wished to migrate to the frontier and "throw myself, at once, into the busiest walks of life," and he thought that he would be either an utter failure or a triumphant success. W. B. Blake longed to go west to find "constant employment" and escape the "lazy life we lead here" in North Carolina. To work alone, to take risks, to strive in a new land—this was the essence of being a man.[12]

Finally, as some of these comments suggest, many migrants had excitable temperaments that set them apart from their peers. Certainly their relatives and friends thought so. A Virginian recalled that his kinsman who left for the Southwest had a "restless unsettled disposition," and a neighbor once described Andrew P. Calhoun, the son of John C. Calhoun, as "impulsive." Louisa Cunningham of South Carolina complained that her nephew

William Yancey "could never rest satisfied in one place two months at a time." Some of these young men may have had what we might call obsessive personalities, like the North Carolinian who was "very full of the Western country," according to a relative, and "seems to think of nothing else."[13]

Something has to be said about the young men, less excitable by temperament and more conservative in outlook, who decided not to migrate. Many of them had absorbed their parents' values, including their attachment to the seaboard and their loyalty to the family at large, and they were more willing to conform to their family's expectations rather than take the risks associated with migration. Furthermore, these men must have reckoned that operating a working plantation, even if it was only moderately productive, was easier than starting fresh in a new country. These psychological and material factors often combined to keep some sons at home.

Henry H. Townes, the brother of Samuel Townes, was typical of the men who decided to remain at home. Born in 1804, the oldest of the five sons of Samuel Townes, Sr., he was a careful, rather somber man who lacked his brother's passions. For instance, when Henry quarreled with Benjamin Perry in 1831, hot-tempered Samuel urged him to challenge Perry to a duel, but Henry declined, and the dispute eventually died down. He took his duty to his relatives seriously, a theme he reiterated in his letters, and he exchanged favors with a range of kinfolk. One of his homes was called Social Hall, appropriately enough. He took an active interest in the lives of his brothers, deluging them with unsolicited advice about their manners, educations, romances, and career choices, and he never ceased urging them to write and visit.

In 1827, Henry Townes married Lucretia Calhoun, a daughter of William Calhoun and niece of John C. Calhoun, a match which anchored him even more solidly in the seaboard. He was evidently a favorite with William Calhoun, who named him an executor of his will in 1840. His connections with the Calhoun clan helped his medical practice, and he ran successful farms in Greenville and Abbeville counties. Henry Townes admitted that the Southwest offered "brilliant" prospects for planting, but he did not wish to migrate. He found life in the seaboard satisfying enough.[14]

Planters' sons who wanted to migrate needed tens of thousands of dollars to purchase a slave force and a plantation in the new country, and they tried to get the capital in a variety of ways. Some tried the time-honored method of looking for a wealthy woman to marry. A South Carolinian told a friend to find "a lady of accomplishments, i.e., worth 100,000," and another urged his unmarried brother to "get all the loot you can." Yet others managed to save some of their own money. William A. Lenoir saved three thousand dollars, to which his father contributed exactly six dollars. Young men might have taken out loans from local banks, but most institutions were reluctant to lend money to individuals with little collateral who planned to move to a distant place; some seaboard banks also limited loans to several thousand dollars per borrower, which was not enough for aspiring planters, or required repayment within several months, which many young men could not do.[15]

Young men could also inherit property, of course. Twenty-two migrants from eleven different families in Table 2 moved after their fathers died. (The exact migration dates for seven other men are unknown.) Six of these fathers left wills, and four (Carr, Hackett, Townes, and Whitfield) divided their estates in a roughly equitable manner among their children. In addition, Peter Field Archer died intestate, so the state of Virginia divided his property equally among the heirs. But the individual legacies in these families were small, ranging from one to seven slaves. Two fathers, North Carolinian William Polk and Virginian John Tayloe III, devised inequitable wills, but because they were so rich their sons who wished to migrate inherited enough property to finance a move to the frontier. Six of the seven Polk sons who lived to adulthood inherited sums ranging from $8,700 to $49,500, enabling them all to move to Tennessee. Tayloe favored his two elder sons over his four younger sons, but Henry, the fifth son, received as many as eighty slaves before he left for Alabama. In the most cold-blooded sense, this was perhaps the ideal solution, since a father's death freed a son of the Oedipal struggle even as it supplied him with property.[16]

Planters' sons who could not raise capital in these ways—by marriage, savings, bank loans, or inheritance—found themselves in a quandary. Their fathers, from whom they wished to establish their independence, remained the best source of capital. So some young men turned to their fathers for help with the start-up costs, the first of many ironies about this generation's search for autonomy. These young men described migration as if it were a right, a matter of individual choice, which was exactly what their fathers would not grant. They thought that migration was a creative act, not the destructive act that their parents feared it would be. For all of these reasons, the dialogue between generations of men was highly charged.[17]

The Fathers

The fathers of prospective migrants were a prosperous, successful, and immobile generation. Most of them were born in the mid- to late-eighteenth century to affluent families who had been in the South for at least a generation, and they came of age, married, and settled down to live as planters in their native states. Despite the economic impact of the War of 1812 and the Panic of 1819, they were an affluent group in the 1820s and 1830s, and many had embarked on second careers in law or politics. They were on the whole so successful that they could not quite understand the difficulties of their sons who came of age by the 1830s. They continued to use conventional agricultural methods, and they had little practical advice to offer their sons when the fields of the seaboard seemed to be worn out. Because of the growing life expectancies of the nineteenth century, many of these men, unlike their own fathers, lived to see most of their sons reach adulthood, and they had to consider the possibility of distributing their resources among their living children.[18]

Many fathers believed that migration to the Southwest was a dubious enterprise. Prices in land and slaves varied a great deal, but a frontier plantation of several hundred fertile acres cost between twenty and thirty thousand dollars in the mid-1830s, and a workforce of twenty slaves cost between eighteen and twenty thousand dollars. Transporting twenty slaves to the new country cost as much as one thousand dollars, and outfitting a plantation with tools and livestock cost five or six thousand dollars. Overseers in the Southwest had to be paid three to four hundred dollars a year, plus bonuses, and land would have to be cleared and homes, barns, slave quarters, and other buildings constructed. To all this would be added the son's living expenses in the new country. The father might have to support the entire ensemble for two or three years before the first substantial harvests came in and the plantation began to pay for itself. In the meantime, his son or his slaves could die from the epidemics of malaria, cholera, and other illnesses that swept the region. The venture could cost over sixty thousand dollars by the time it turned a profit. Even if a son contributed toward these costs or a kinsman or friend helped defray them, it would still be an expensive project.[19]

It seemed much more sensible for a son to stay at home. A father would probably pay less for a cotton plantation and slave force in the seaboard—although once again, prices varied—and he would be spared the other expenses of establishing a new concern on the frontier. He could also monitor his son's management of any resources he gave him and benefit in his old age from his son's support. Historians estimate that a nineteenth-century father began to receive a return on his financial "investment" in raising a son when the child reached age twenty-seven. If this was the case for planter fathers, then the migration of a son in his mid- to late twenties would understandably be resisted. Planter fathers did not describe their sons as investments, of course, but they clearly expected their sons to remain near home to help with their affairs and, if necessary, care for their dependents after they were dead; this was part of the generational pact. William H. Brodnax, who raised his younger brother and looked upon him as a son, remarked, "He has more than repaid in later years my earlier services." Brodnax expected his younger brother to "assist in protecting and raising my little ones" after he was gone.[20]

Fathers resisted the migration of their sons for more deeply personal reasons. They grew up in the eighteenth century, the age of "agrarian patriarchy," and as adults they believed that they had the right to take an active part in the decisions of their sons or even make decisions for them. They also wanted to preserve their status by commanding the obedience of their dependents. Preventing migration was a show of power; conversely, a son's migration was a blow to the father's power.[21] Furthermore, planter fathers had their own ideas about masculinity. They believed a man should place the family's welfare above personal goals. A responsible man, and a good son, was loyal to the welfare of the entire family.[22]

The standards of masculinity proposed by the younger generation therefore bewildered and angered many older men. A North Carolinian believed that his nephew, who wished to go the Southwest, lacked "the will to be a man." He observed that the spirit of the times was "unfavorable to the growth and development of the manly virtues." Instead young men indulged in "idleness and extravagance." He lamented, "Industry and economy, old guides, and venerable land marks, are now generally repudiated, and considered as belonging to another, and darker age." Although James Henry Hammond had himself considered migrating as a young man, he was exasperated by his sons' aspirations as they reached adulthood in the 1850s. He thought their wishes for individual autonomy were a "ridiculous fallacy," exclaiming, "Who ever was, who ever can be independent?"[23]

Many fathers exercised their authority and by sheer force of will made their sons stay at home. Kenneth McKenzie ordered his stepson to remain in North Carolina because of the proximity of relatives, cautioning him, "Don't give up your home for a song," and then stating flatly, "I insist." In some families a father's authority lasted throughout his life, inhibiting migration long after sons became adults. One man admitted that he could not migrate during his father's lifetime but hoped to move to the Southwest after his father's death—if he survived him. Bushrod Powell reached from beyond the grave to prevent his offspring from moving. He inserted a clause in his will that required his brother-in-law's approval before his estate could be used to help his sons migrate from Virginia.[24]

Others decided to bargain with their sons to keep them from migrating. John J. Ambler, Sr., knew by the mid-1830s that several of his sons were considering moving from Virginia to the Southwest, in part because they were unhappy with the bequests he planned to leave them in his will; he was in his early seventies, and he needed someone to help him run his various enterprises. He decided to make a generous proposition to his youngest son, William, just as he was preparing to leave for Alabama in 1835. Ambler offered his son twenty slaves, the profits of a mill he owned, and one hundred dollars in cash if William would remain and attend to the family's business. William had long been chafing at the disadvantages of his birth order, but he finally accepted on the grounds that his duty to his aged father and the property he would instantly acquire outweighed the "fortune" and "eminence" he could win on the frontier. He stayed in Virginia for the rest of his life.[25]

A father's disapproval was not always insurmountable, however, as Langdon Cheves discovered. A rich planter, politician, and banker from South Carolina, Cheves was able to prevent his son-in-law Thomas P. Huger, husband of Anna Cheves Huger, from going to the Southwest, but he could not stop his son Alexander from leaving. Cheves gave the Huger couple a plantation in Abbeville County and slaves that were worth some thirty-five thousand dollars, but Thomas Huger soon complained that the concern was not turning a profit and that he was living in a state of "dependency," which he was "determined to avoid." When he said that he wanted to migrate, Cheves roared back, "If I have the power, I will resist it." He did have the power: he insisted that Huger was just mismanaging the place and forced him to remain on it.

But somehow Cheves could not prevent Alexander, his second son, from leaving for Alabama in 1841. The relationship between father and son had long been a troubled one; Cheves said that he had "suffered" from his son's "unfilial conduct" since Alexander was a boy. He gave Alexander at least twenty-one thousand dollars from his inheritance, but this was not enough to enforce his son's obedience. Alexander left home and apparently never visited or wrote to his family before he died a few years later from the complications of alcoholism.[26]

Other men watched their sons migrate without their permission, and they felt very much betrayed. David Leech was devastated when the last of his five sons left his South Carolina home for the Southwest. "This is the way my children hath served me," he exclaimed, "after all the care and pain I have taken to secure to them my plantation"; they "left me to shift for myself in old age they are all gone." Most insulting of all, one son sold land his father had given him and used the money to finance his trip to the frontier. When Leech wrote his will ten years later, he left nothing to his sons. His neighbors in Chester County, South Carolina, where Leech died, still recalled that his sons had deserted him in this "unnatural" way.[27]

A minority of fathers agreed to help finance the venture, however, and they bear further examination. Before William Polk died he advanced some $44,500 in cash, land, and slaves to his son Lucius, including a plantation in Maury County, Tennessee, but he also forced Lucius to go there and manage the place, even though his son did not enjoy farming. John J. Ambler, Sr., gave about eleven slaves each to his sons John and Richard but apparently decided later to deduct the equivalent sums from their inheritance. Both actions represent yet other exercises of patriarchal power.[28]

A more interesting minority helped their sons with little or no expectation of repayment. North Carolinian William Lea provided generously for all of his sons throughout his life, giving or lending them thousands of dollars' worth of property whether they settled in North Carolina or fanned out across the Southwest. When Andrew P. Calhoun wanted to leave South Carolina for Alabama in 1838, his father John C. Calhoun, lent him three thousand dollars with interest and persuaded his son-in-law Thomas G. Clemson to lend Andrew another seventeen thousand dollars. (John Calhoun's cousin Francis Pickens thought it was a terrible mistake to invest so heavily in "a new and uncertain country" at such a "great distance" from home.) These loans became a gift, in effect, because Andrew never repaid the sums. Clemson, a Northerner, grew more infuriated as the years went by, while John Calhoun waited patiently and told his son-in-law to do the same, reminding him that he ought to trust a business partner who was so "intimately connected" to him.

The social and economic backgrounds of Lea and Calhoun resembled those of others in their generation; they were involved in the usual patterns of economic assistance with their male relatives, and both prospered in the seaboard. Yet both departed from the standards of fatherhood and masculinity typical of their generation and took a less authoritarian view of their paternal duties. They may have felt so secure in their accomplishments that they did not believe their authority was at issue. Lea was a man of affairs in Caswell County, North Carolina, a merchant and minister as well as a planter. Calhoun moved on a much larger stage, of course; he was a powerful politician, a national figure, and an immensely self-confident individual in his own abstract way, so much so that he may have welcomed signs of ambition in his firstborn. Furthermore, as early as 1820 he had a positive view of the Southwest, telling a friend who had migrated to Alabama that "the mere inconveniences of a new country will soon wear away while the invigorated hopes, which it usually excites, will long continue." But these men were not representative of their generation. The majority of planter fathers in the families in this study did not assist their sons.[29]

Wives, Mothers, Sisters, and Daughters

The differences between the sexes on the migration issue were even more pronounced than those between generations of men. Women, with few exceptions, feared and dreaded the possibility of moving to the Southwest. Migration could not represent independence for them, as it did for young men, because women spent their entire lives in a state of dependency. If it came during a woman's childbearing years, it would deprive her of the practical aid of relatives in running households, and it would deprive her offspring of relationships with kinfolk that were crucial to socializing children. If a woman had to migrate in her old age, she would spend her last years far from relatives who would care for her. Finally, it would separate all women from the female kin who gave them love, companionship, and a sense of identity throughout their lives.

Women accepted the partial separations from relatives that happened when they married, but migration portended losses that were radical and permanent because it could remove women so far from their kinfolk that those relationships might deteriorate and eventually collapse. The frequent visits that kept kinship ties alive in the seaboard would be improbable, maybe impossible, from a new home hundreds of miles away. Journeys over great distances were unsafe and could not be undertaken without the company of a man, and many women feared such journeys. For example, widow Patience Laye of South Carolina decided to visit her daughter and son-in-law in Florida only "if you was to come . . . and help me there."[30]

Ideally, many women wished that their relatives would remain close enough for daily contact. One North Carolinian objected when her brothers moved twenty miles from home, just beyond a day's ride on horseback, and she told her cousin that the separation "of course will render us somewhat gloomy every day." Migration beyond the boundaries of a state was troubling, but the distance from the seaboard to the Southwest seemed unsurpassable. When the cousins of Catherine R. Patterson left North Carolina for Mississippi, she believed that she would not see them again in this world and pinned her hopes on a reunion in heaven. A distance of two hundred miles, according to Caroline Gordon, made visits highly unlikely. If this distance was a threshold in the minds of women, then they would certainly be anxious about removal to the Southwest, since the average distance per migration in Table 2 was approximately five hundred miles.[31]

Women commonly compared migration to death, as men almost never did. A North Carolinian walked out into her yard to take a long "final look" at her sister as she left with her husband for Missouri in 1835. Another woman ran after her daughter as she departed with her husband, embraced her one last time, and cried, "Oh, Mary, I will never see you again on earth." Women compared the two losses so frequently that it is sometimes impossible to tell if "departed" relatives had died or migrated. Louisa Cunningham mourned the loss of her beloved sister whom "providence has then seen fit to remove away from us." It was "a circumstance at all times melancholy and distressing—sundering apart those near and dear ties which so long has bound [us] together." To many women, family relationships were little short of life itself. Perhaps it was no accident that the word "removal" was used for both migration and death.[32]

Planter women described the dispersal of their relatives as if it were unprecedented, as if these upheavals followed a long period of tranquility. While the Southern white population as a whole was highly mobile throughout the antebellum era, the families of some planter women actually had not moved in many decades, and these women demonstrated an abiding love for their homes. The family of Ann Gordon Finley had lived in Wilkes County, North Carolina, from the time of the Revolution until the 1840s. She was born in 1826, the third child of Sarah Gwyn Gordon and Nathaniel Gordon, and her father died in 1829. After attending Salem Academy, she was married at age twenty to John T. Finley, a local slaveowner and merchant. She was a practical, down-to-earth woman, but she was also closely attached to her many kinfolk from the Gwyn, Gordon, and Lenoir families, perhaps more so than the typical planter woman.

Ann Finley was dismayed when her husband planned to move to Alabama in 1847 because, as her mother said later, "Ann was so fond of her relations." She also loved the hilly landscape of western North Carolina, especially her childhood home, the Oakland plantation located near Wilkesboro. Before her departure Mrs. Finley visited her mother there and was saddened by the air of desertion about the place. "Once the habitation of so large a family," it was now empty except for her mother and stepfather. It had been "so long the homestead of the Gordon name" but "there was not one to answer by that name."[33]

Many women did not want to move because they perceived the Southwest as a violent, dangerous, sickly place. Like men, they constructed their images of the frontier from their reading and from conversations with relatives and friends. A Virginian recounted the grisly deaths of a neighboring family who had gone to Texas only to be murdered by a gang of thieves. Olive Packard worried that the Southwestern climate would ruin the health of her brother and sister-in-law, and Rachel Townes feared for the health of her children, Samuel Townes and Eliza Townes Blassingame, in Alabama. Furthermore, women believed that frontier life would promote "dissipation," meaning drinking, gambling, and miscegenation. Somewhat inconsistently, they also dreaded its isolation. Mary Boykin Chesnut's visits to the Southwest over several decades left her with an impression of dreary loneliness which she recalled as "despairing" and "depressing."[34]

Some women even criticized the greed of men who went to the Southwest. Flinty, plain-spoken Anne Dent chastised her son for the avarice that drove him from South Carolina to Alabama in 1837. Had he been contented with "moderate gains," she scolded, "we might yet have been enjoying family union, instead of being so widely scattered and unavailingly lamenting our mistake." Frances Berkeley of Virginia remarked acidly that one of her neighbors was "much pleased at first at the prospect of making a handsome speculation," but the man instead died from fever in Tennessee far from his wife and children. After Lydia Riddick's neighbor moved to the Southwest despite his mother's protests, she told her own son, who had also migrated, that "'tis strange that the love of money is stronger than the love of a kind and affectionate mother whose heart yearned to look on him." These women spoke directly to the issue of what constituted acceptable sacrifices in the pursuit of wealth, and they believed that family life was more important than material gain. Women viewed migration primarily as a moral issue, unlike either the young men who saw it as an opportunity to fulfill individual goals, or older men who saw it as a challenge to their power.[35]

Women nonetheless played only a marginal role, at best, in decision-making. Men knew that they opposed migration (Willis Lea advised his brother to leave North Carolina although "all do not like that move to the west—especially ladies"), but men almost always discounted women's perspectives. James Lide moved from South Carolina to Alabama over the strong objections of his adult daughters. Israel Pickens realized that his wife, Martha, did not want to leave North Carolina for Alabama, but he decided to go anyway, teasing her about her fears of the wilderness.[36]

Most planter men did not consult their wives, in either new or long-standing marriages. One husband brought his bride to Alabama "away from her family and friends, to a land of strangers, where she finds much to regret." After seven years of marriage, Elizabeth B. Ambler, wife of John J. Ambler, Jr., tried to dissuade him from buying a plantation in Alabama. She begged him not to settle there—"*Heaven forbid,*" she exclaimed—but he chose nonetheless to relocate in the Southwest. After twenty-three years of married life, North Carolinian Sarah Gordon Brown was not included in her husband's deliberations. She told her son, "He sais [*sic*] he is going to move but makes no farthere [*sic*] preparation and none of us know what he is going to do."[37]

Mothers also spoke out against migration, but with little effect. Elizabeth Otey objected that her son's migration from Virginia to Tennessee was a source of "great grief" and urged him to return from the corruptions of the frontier. Two years later she wrote that his hopes for success had been unrealistic and it was time to come home. After Sarah Irby's son departed Virginia for Mississippi, she regretted it more each day, wailing, "Would to God you had never gone." When Mary Allen's son left home, she was hurt and dismayed, asking, "Why did you go to Texas?" The next year she wrote that she could never be reconciled to his absence. Yet none of these men returned to the seaboard.[38]

Sometimes women could convince men to postpone their departure if they had the backing of their male relatives. A former slave recalled that her master's in-laws "riz up an' put forth mighty powerful objections," persuading him to delay leaving South Carolina for the frontier. William Townes, an uncle of Samuel Townes, heeded the wishes of his wife and children who very much wanted to remain in their native Virginia and searched for a suitable purchase in the Old Dominion. When he could find no promising land in the state, he finally decided to leave for the Southwest. Husbands sometimes attempted to alter their wives' opinions of the region. Samuel Van Wyck tried to depict Huntsville, Alabama, as a place fit for "a lady and a Methodist" like his wife.[39]

A handful of women went willingly to the Southwest because they wanted to escape home. The Townes family opposed Samuel's marriage to Joanna Hall, perhaps because her family was not wealthy; whatever the reason, she was relieved to escape their disapproval and, more generally, the intense family life of the seaboard. According to her husband, she was glad to have a home of her own in Alabama. Elizabeth Witherspoon DuBose, wife of Kimbrough DuBose and part of the enormous DuBose-Witherspoon-Miller clan, gave up her initial opposition to migration after living for ten years in a kinsman's home in Cheraw, South Carolina. Her son related that she too decided she wanted a home of her own.[40]

At least two women migrated to escape unhappy marriages: a South Carolinian took refuge with her Alabama cousins, and a Virginian moved to Kentucky to live with her sons. But these were only temporary solutions to the problem of domestic infelicity, and both women eventually returned to their homes. Women, unlike men, could not escape the dilemmas of familial obligation by settling alone on the frontier.[41]

Despite their general opposition to migration, most women struggled to accept the decisions of their male relatives. One mother wished her son would return from Texas, but if he chose to stay, then "of course I ought not to object." Samuel Townes predicted that his female relatives would soon reconcile themselves to his sister's and brother-in-law's departure for Alabama. Eight months later they were still trying to accept the separation, but no one openly challenged the decision to leave Carolina. The dependence of women was the counterpart to the independence of men, and at some point most women had to at least feign agreement with the decisions of their male kinfolk. Bernard Cox summed up the expectations of many men when he announced his engagement to a woman who was "willing to do as I would have her do"—accompany him without complaint to the Southwest.[42]

The Slaves

The true captives of migration were, of course, the slaves. Between six hundred thousand and a million slaves moved from the seaboard to the Southwest in the antebellum era, and the majority went in the 1830s.[43] Seaboard slaves had already formed highly negative images of the frontier, created, perhaps, from informal communication networks within the slave community. They feared the harsh working conditions and climate of the Southwest and, most terrible of all, the destruction of family ties, which would be even more complete than what planter women feared would happen to their own families. Slaves who went to the Southwest could not look forward to regular letters from their relatives in the seaboard, nor could they return home for visits at will. As Frederick Douglass wrote, the "removal" of a slave really was equivalent to death.[44]

An examination of sex ratios among seaboard slaves indicates that planters probably did not breed slaves for sale or systematically sell large numbers of slaves of either sex, as some historians have alleged. The sex ratios for slaves between the ages of ten and fifty-five for six counties in Virginia, North Carolina, and South Carolina fluctuated through the census returns from 1820, 1830, and 1840, with some counties showing males in the distinct minority, while others had a majority of males. (See Table 3.) All six counties showed decreases in sex ratios between 1830 and 1840, when most slaves went to the Southwest, but none of them witnessed the sharp declines that would be associated with massive sales of men to the frontier. These numbers are far from conclusive, but they suggest that slave men slightly outnumbered slave women in the population that went to the Southwest.[45]

Planter men nonetheless hardly ever mentioned family ties within the slave population as they prepared to go west. Individual planters often bought and sold slaves before moving west, separating relatives from each other. Even slave families that went to the Southwest with their masters lost family members because they had relatives on other plantations, and, as other scholars observe, slaves counted many relatives beyond the nuclear family as significant kinfolk. So even if a master did not sell a single slave, he destroyed vital links between relatives and undermined the slave community when he moved to the Southwest.[46] Other slaves went to the Southwest even when their masters remained in the seaboard. Some planters sold rebellious slaves (both men and women) to the frontier to punish them, while others sold mulattoes whose physical resemblance to their white fathers was too obvious. As Orlando Patterson points out, even if these partings occurred only occasionally, the possibility frightened many slaves.[47]

Most slaves were helpless to stop these separations, but one man was able to reunite with his children after they were sold to the Southwest. Charles Ingram ran away from his master in Richmond, Virginia, in the late 1840s or early 1850s and found work outside the city, somehow passing himself off as a free man; meanwhile, his wife died and his sons remained in bondage. But when Ingram discovered that his sons had been sold to Texas, he pursued them to the frontier and agreed to live as a slave again in order to be with them. This dramatic account not only reveals one man's profound love for his offspring, but also suggests the anguish that migration created in thousands of slave families whose members had no control over their fate.[48]

Planter women were usually more aware of the fears of migrating slaves than were planter men, and some disapproved of the division of slave families. One married couple argued over the issue, according to a relative: "A good deal of trouble in Nan's family—because of Tom's insisting upon a sale of some of her negroes." Women missed individual slaves who went to the Southwest and inquired about them, something that men rarely did. Rachel Townes sent greetings to "*old Lucy particularly*" after her son Samuel took the slave to Alabama with him. Elizabeth B. Ambler, who remained in Virginia, asked her husband, John J. Ambler, Jr., to return Adam, her favorite slave, before he was stricken by an epidemic sweeping their Alabama plantation; she requested the names of slaves who had already died, which her spouse neglected to mention in his letters. Widows who owned bondsmen took concrete steps to protect their slaves. Ann Macrae stated in her will that her slaves should not be sold out of the county without their permission and should not be subject to the control of her sons-in-law.[49]

Change Over Time

In the years after 1840 many continuities and some changes characterized the decision-making process of seaboard families. Young white men continued to see migration as the way to achieve "manly independence." When Samuel Van Wyck planned his departure from South Carolina for the Southwest in 1859, he described his goals in language similar to that used by young men in the 1830s. He longed to take his place on "the world's busy battle ground—to meet opposition—to trample it underfoot," proclaiming, "I feel that the West is the land for me." Until the end of the antebellum era, the Southwest functioned as a symbol of opportunity and liberation from the confinement of the seaboard. But the decision to migrate remained a dialogue between planter men, excluding planter women. In 1859 Philip P. Dandridge told his twin sister, who was his best friend and confidante, that he had decided to go ahead with his plans to migrate to Texas although he knew that she opposed them. Slaves continued to be shut out of the decision, of course, and many went to the frontier unwillingly. Issabella Boyd, a former slave, recalled her feelings as she traveled from Virginia to Texas in the 1850s: "Every time we look back and think 'bout home it make us sad."[50]

The seaboard economy began to change, however, after 1840. In Virginia farmers began to diversify crop production, and after 1840 its outpouring of migrants began to slow down; new settlers from the Middle Atlantic and Northeast began to move into the state. The flow of emigration from North Carolina also decreased, but the economy never fully recovered from the hemorrhaging of the

1830s and the recession of the 1840s. The economy of South Carolina, still tied to cotton production, improved in the late 1840s and 1850s as prices of the staple increased. Despite brighter prospects in the seaboard, however, the drain of migrants from the region continued at a somewhat slower pace for the rest of the antebellum era.[51]

Throughout the period, planters' sons entered into a national process of myth-making about the American West, projecting their needs and wishes onto the Southern frontier. Planter women, like many other American women, looked upon westward migration as an "antimythic" experience, one of separation and loss. Migration took on antimythic stature among slaves, too, as a tale of grief. Once the decision to migrate had been made, planter men, planter women, and slaves departed the seaboard with dissimilar expectations of life on the frontier.[52]

3

A New World: Journey and Settlement

From the moment planter men and women said goodbye to the seaboard, they experienced the journey and the early phases of settlement in different ways, reflecting their contrasting resources, how much autonomy they enjoyed, and the different roles they played in the family. Men retained their decision-making power, and the choices they made, such as where they would live, reflected their goal of freeing themselves from the family. Therefore they saw the landscape of the new country as attractive, open, and abounding with opportunity. Women, by contrast, mourned the severance of family ties, and they experienced the landscape as depopulated, bereft of kinfolk. The slaves who went with the planter family also grieved for the relatives they left behind, perhaps even more intensely than whites.

Furthermore, everyone had to confront a physical environment that differed considerably from that of the seaboard—although, once again, planter men, planter women, and slaves experienced these changes in diverse ways. Everyone worked hard, harder than at home, but the duties of slaves and planter women changed more drastically than those of the white men who brought them to the frontier. Yet even planter men discovered that the Southwest was not always a land of promise.

Departure

Most people realized that migration was a turning point in the lives of everyone in the family, and they remembered the day of departure long afterwards. Those who stayed behind felt great sorrow. When Kimbrough DuBose left South Carolina for Alabama, his father-in-law, John D. Witherspoon, who was seventy-two years old, "wept bitterly" when he said goodbye to his daughter and nine grandchildren on a "dark drizzly morn." As Samuel and Mary Maverick departed for Texas, Samuel's father rode with them for half a day, and, according to Mary Maverick, he was filled with "grief when he at last parted with his son." Everyone in the Blair family of Virginia felt "tears and pangs" when they gathered to see one of their sons off to Texas.[1]

Some young men also felt the significance of the occasion. Abner Grigsby remembered "the thrilling grasp—the fond embrace—the parting glance" of his relatives as he rode off one November morning for Texas. But many left with few regrets; they were eager, even desperate, to escape the family and the seaboard. One planter refused to make a farewell visit to his brother's plantation because it would delay his departure and cost him money. John J. Allen, Jr., left Virginia abruptly in the wake of a dispute with his father; his mother wrote that they both regretted "your having gone to Texas in such a way and with a stranger." Some did not even bother with goodbyes. Three cousins simply disappeared from their homes in North Carolina, stunning their relatives and friends.[2]

Planter women, however, were often distraught as they set out for the frontier. On the eve of her departure for Alabama, Mary Ann Taylor told a friend, "you *cannot* imagine the state of despair that I am in." Jane Lide Coker recalled the profuse weeping among her kinfolk when her family left for Alabama in the fall of 1835. "It was heart-breaking," she wrote, "for in those days the distance seemed vast." Saying goodbye to female relatives was especially painful, and these farewell scenes made a lasting impression on women. After Evalina Lenoir left North Carolina with her husband in 1835, she vividly recalled "dear sister Nancy standing in her little porch catching the last glimpse and dear sister Louisa with all the girls taking the last look."[3]

The most searing farewells took place in the slave quarters, because those who watched their loved ones go realized that this was probably a final goodbye. When Laura Clark left for the Southwest as a small girl, her mother, Rachel Powell, ran after the wagon, fell to the ground, and "roll over on de groun' jes' acryin'"; it proved to be the last time Clark saw her mother. Most slaves knew little of geography, which must have made these separations all the more harrowing. Binah DuBose, whose son Ned had been taken to Alabama in the 1830s, knew that he belonged to a planter named Pegues in Dallas County, but, as one of her masters recalled, she had "no conception" of where it was located. Those who departed often literally did not know where they were going, but they sensed that they were going a "'returnless distance,'" as one white man remarked.[4]

Sometimes slaves did not even know who would be leaving until the very day of departure, as demonstrated by the chilling events on the plantation of Kimbrough DuBose, who left for Alabama with his wife, children, and slaves in 1850. DuBose was deeply in debt, and the county sheriff appeared that afternoon demanding that he pay a creditor the sum of $1,650. Then his brother-in-law, planter John N. Williams, who had come to say goodbye, volunteered to pay the debt if DuBose would leave three young male slaves as collateral. He hastily agreed, choosing from his assembled slaves three brothers, Sandy, Jake, and Steve, whom he gave to Williams. DuBose's son, John, who recounted the incident, did not mention the boys' reactions or those of their families. This too proved to be a final goodbye. DuBose never repaid the debt, and the brothers remained in South Carolina on the Williams plantation.[5]

The Journey

Even though men and women, whites and blacks, traveled together to the new land, they experienced the journey in diverse ways. For planter men, this trip resembled others they had taken in their lives, such as going away to college for the first time, in that it involved linear movement through geographic space to a fresh destination, but planter women for the most part had experienced mobility as a cyclical movement from one familiar place to another with an eventual return to their homes. For them, a journey through the strange landscape of the Southwest to an unknown destination was a novel and disturbing experience.

Some slaves had accompanied their masters on trips through the seaboard, and others had visited plantations near their homes. Many had experienced forced migration when previous masters sold them to new owners. A minority, mostly young single men, had traveled to other destinations when they ran away from their masters, only to be recaptured and returned to bondage. Many slaves knew their broader neighborhoods quite well, but the great majority had never traveled as freely as planter men, nor had they experienced the rhythmic mobility of planter women. So their odyssey was both familiar and unfamiliar: once again they were going somewhere at the bidding of white people, but they were headed for a destination that filled them with foreboding.[6]

Most of the planter men in this study departed in the 1830s, like most whites who left the seaboard for the Southwest in the antebellum era.[7] Some single men went alone to explore the Southwest, but others attached themselves to families, and married men, in turn, wanted other men to travel with them for safety's sake. When Thomas Brown left Virginia for Florida, he set out with sixty slaves and "about twenty young [white] men, who desired to adventure with me."[8]

Planters usually departed in the fall, when roads were hard and dry and rivers had subsided from summertime levels. Most traveled by horseback or by wagon, and most followed the main roads, such as the Cumberland Road, which ran through Tennessee, the Federal Road, which ran from Georgia to Natchez, Mississippi, or the Natchez Trace, which began in Nashville, Tennessee, and also ended in Natchez, although some parties took other, less well traveled routes.[9]

The typical journey from the seaboard to the Southwest was approximately five hundred miles in length, that is, from a Piedmont county in North Carolina to the Black Belt of Alabama (named for the color of its soil), the most common starting point and destination for the planter migrants in this study. This was no easy undertaking in the antebellum era. Even the principal roads were poorly maintained and badly marked, narrowing in some places to little more than trails in the wilderness. Parties of migrants occasionally stayed at taverns or with relatives who had already settled in the Southwest, but most nights they camped outdoors, sleeping in immense forests where panthers, bears, and wolves roamed.[10]

The journey could be time-consuming and costly. A man traveling alone by horse could make the quickest time, advancing as many as twenty-five miles a day, but larger parties moved about fifteen miles daily. Most migrants were on the road for at least a month. A single man could spend between one hundred and two hundred dollars for a trip in the 1830s, and expenses for a white family and twenty slaves could exceed one thousand dollars.[11]

Planter men enjoyed moving through new landscapes, relishing the strangeness, the stir, and adventure. As Richard C. Ambler traveled to Alabama, he exclaimed that "the number of emigrants surpasses all calculations. . . . For six or eight miles at a time you find an uninterrupted line of walkers, wagons, and carriages." Thomas Brown enjoyed his two-month odyssey to Florida so much that he would have been content "to be a wandering Arab the rest of my life." In these caravans men selected the route and led the entire party down the road. They chose the resting place for the night, helped set up tents, and then relaxed while planter women and slaves cooked and cared for the children. A few men felt saddened by the upheaval, such as the North Carolinian who had "the blues" on his way to Louisiana, but for many men the journey itself was an affirmation of masculinity and independence.[12]

Planter women, by contrast, found the journey to be an unhappy, even a tragic, experience. Unlike men, they never traveled alone to the Southwest but went in the company of male relatives or close family friends. They rode with their children and female kin in carriages, while wagons and pack horses carried the family's belongings.[13] The presence of kinfolk made the trip less lonely, but women still mourned for those they had left behind, as did Sarah Jane Lide Fountain, who cried that "my heart bleeds within me" whenever she thought of the "many tender cords that are now severed forever." Women also found the natural environment frightening. Many worried about exposure to sickness, and others feared the wildlife in the forest. A former slave who made the westward journey with her owners recalled that "de missus, she wuz scared at night." Some women were pregnant during the trip, which often increased their physical discomfort and at the same time heightened their sense of vulnerability. Frances D. Polk, the wife of Leonidas Polk, was ill during the entire journey from North Carolina to Tennessee in 1833, but the party pressed forward; she gave birth to a stillborn child before the family reached its new home in Maury County.[14]

The longer the trip, the more disruptive it may have been for planter women. When Jane Woodruff, a demure, sheltered woman from Charleston, set out on January 15, 1823, for Florida with her husband, Joseph, their three children, and several dozen slaves, her heart quailed at the prospect of leaving her "native country." A terrifying journey lay ahead of her. The entire party nearly drowned while crossing a river before they left South Carolina. After travel over what Mrs. Woodruff called "dreadful" roads, the family reached St. Mary's, Georgia, where she contracted dysentery, which "reduced me almost to a skeleton." She remained there with the children for several months while her husband took the slaves to Florida. Rumors reached the town that he had died in the wilderness, but Mr. Woodruff reappeared, alive and well, and escorted his family to the St. John's River, on which they traveled by boat for fifteen days. The vessel had no sleeping quarters, so the family slept on deck in "excessively cold" weather. The boat ran aground six miles from their destination, so they walked to their new plantation, arriving in February 1824, over a year after leaving Charleston. Jane Woodruff was physically and emotionally exhausted, but she went to work right away feeding the family's slaves. The westward journey taxed the strength of some planter men, but they left no accounts of the trip quite like those of women. All in all, the journey reminded planter women with special force that they had little authority in their families and little control over their destinies.[15]

For slaves the journey was a profoundly degrading experience. Long after Emancipation many blacks were still outraged by their memories of slaves being driven down a road as if they were livestock, and one black woman, Jane Sutton, said that slaves were "stole" from their seaboard homes. The trip was dangerous as well as humiliating; slaves drowned while crossing rivers, and others died of illnesses contracted during the journey. Those families which had not been separated in the seaboard were still at risk because masters sometimes decided to sell slaves to whites they met along the way. Most bondsmen traveled on foot, walking alongside the wagons, chained or tied together, while white men rode close by to prevent their escape. Some slaves were bound together, in fact, because whites knew that they were grieving for family members and might run away. Planters permitted a few slaves, usually house slaves, small children, the sick, or the aged, to ride in wagons.[16]

Some slaves continued to work during the trip, however, driving teams of horses, erecting tents, managing livestock, and preparing meals every evening. They performed other, more personal services for their owners. In one caravan, house slaves helped whites alight from their carriage whenever it stopped. Another bondsman accompanied the small sons of his master so he could lift them up when their boots stuck in the mud. Slave women may have found the journey most disruptive of all. Days of forced marching exhausted them, and pregnant slave women, like pregnant planter women, suffered during the trip. One slave woman who was evidently already in physical distress miscarried after her master, Richard Ambler, insisted that she walk rather than ride a horse.[17]

Destination

Just as planter men decided that families would leave home, they decided where families would settle. Some purchased land in advance, others made a choice after reaching the frontier, and yet others changed their minds along the way. Men with training in medicine or the law collected information about their colleagues,[18] which sometimes influenced their decisions, but most migrants were primarily interested in fertile land that was lightly settled and located near a river or road so they could transport crops to market. They also searched for an area that had a reputation, at least, for being free from the worst outbreaks of fever or sickness. All were seeking an intangible quality called "opportunity," the sense that an area had a bright future and was attracting enterprising, able individuals. Most flocked to the Black Belt of Alabama, the rich counties of central Tennessee, or various locations in Mississippi.[19]

Most young men tended to settle in areas where none or few of their relatives lived, trying to free themselves from what one man called the family's "assistance and control." Hugh Thomas Brown left Wilkes County, North Carolina, which was heavily populated with his kinfolk, for Van Buren, Arkansas, where he did not know a soul. Another planter who had grown up surrounded by relatives moved to Arkansas, where only his cousin's daughter and son-in-law owned a plantation. In a popular area such as the Alabama Black Belt, some kinfolk inevitably settled near each other because they were all drawn to its rich soil, but the number of relatives was still much smaller than the crush of kinfolk at home. An uncle and a brother-in-law of Samuel Townes already lived in Perry County, Alabama, when he arrived in 1834, a mere handful compared to his relatives in South Carolina. Townes remained in the county despite the presence of his brother-in-law William Blassingame, whom he grew to hate with a passion.[20]

Even when men remained on good terms with their kinfolk, they tried to have minimal contact with them. Richard Hackett bought some land near Shreveport, Louisiana, where his uncle and several cousins lived, but he visited them only occasionally, and he lived alone. He told a brother at home in North Carolina that he was determined to remain "independent" of these kinfolk. Josiah Nott settled near his father-in-law in Spring Hill, Alabama, but he did not want to have to visit a "large circle" as he felt obliged to do in South Carolina. Another man who located in Scott County, Mississippi, told his brother that he lived a "good distance" (in every sense) from his relatives, so he did not see them regularly and heard from them only sporadically.[21]

A few men consulted their female relatives about where they should settle, and women of course wanted to live near some of their kinfolk. When one man brought his bride to Mississippi, he may have tried to cushion her departure from home by choosing a state where her sister already lived. But men chose homes according to several criteria, and they were most concerned with the economic prospects of an area; the presence of kinfolk alone was insufficient reason to select a location. Most men excluded women from this decision much as they excluded them from the original choice to migrate. Israel Pickens, who moved to Alabama's Black Belt, callously remarked that his wife, Martha, was "as well satisfied as could be expected at such a distance from all her relations and the acquaintances of her youth."[22]

Settlement

All the new arrivals, men and women, slave and free, had to confront a new physical environment, but they had separate challenges to meet and overcome. Most planter men recognized that the settlement period would have its trials. Parts of the Southwest were still a wilderness, and even in more settled areas some land had to be cleared, buildings constructed, and crops planted. Professionals had to establish practices and reputations. Several years of hard work lay ahead before migrants might begin to reap a profit.[23]

When migrants purchased farms and plantations, they entered an unstable land market. The federal government sold land appropriated from various Indian tribes, and some planters, such as men in the Whitfield family, took advantage of the low prices. At land offices throughout the Southwest, lots of 640 acres sold for as little as two dollars an acre. Good land in Alabama's Black Belt was more expensive, averaging about forty dollars an acre, and prices rocketed as high as one hundred and fifty dollars an acre as settlers poured into an area.[24]

Planters looked for large tracts of land, and they tried to buy as many slaves as they could afford. Cotton prices soared in the early 1830s, from eight cents a pound in 1830 to eighteen cents a pound by 1834 (the crop year of 1833–1834 was the first year of the "cotton boom"), and many men believed that they would soon be rich. They wanted farms that were at least several hundred acres in size, but many bought much larger tracts, perhaps because they were striving for the economies of scale possible on a large plantation. Henry Tayloe of Virginia purchased 1,620 acres in Alabama in 1834 and then bought hundreds of additional acres in Marengo and Perry counties; Andrew P. Calhoun bought 1,240 acres in Marengo County in the 1830s. Some planters brought slaves with them, and others continued to buy and sell slaves after they arrived on the frontier. They purchased bondsmen from local slave traders or other planters, while yet others returned to the seaboard to purchase slaves. Prices for young male slaves rose, too, from several hundred dollars in the early 1830s to as high as $1,200 in 1837. Prices of just about everything shot up every day, according to Gaius Whitfield.[25]

Settlers paid for land and slaves in a variety of ways. A few received familial assistance, as did Andrew P. Calhoun, and some, such as William A. Lenoir, drew on their personal savings. Henry Tayloe sold some of the property he had inherited to finance purchases in the Southwest. He did so well that he began to lend money to other settlers at exorbitant interest rates, as high as 30 percent compounded every six months.[26]

To these and other moneylenders many young men turned if they got little or no assistance from their families. They bought property with bank loans, which were easy to get before the Crash of 1837. President Andrew Jackson vetoed the second charter of the Bank of the United States in 1832 and then distributed government funds in a number of state banks, the "pet banks." The federal government did little to regulate the nation's banks, however, so by the middle of the decade badly run institutions proliferated all over the United States. The entire country witnessed an inflation in prices as specie from abroad, primarily from Mexico, flooded the country. These national trends are visible in the "flush times" on the Southern frontier, as banks and wealthy individuals lent thousands of dollars to settlers with little collateral in land, slaves, or crops, setting off a frenzy of buying. Samuel Townes came to Alabama with only three slaves, but he boasted, "I can command credit here for *any* amount of *property*." Many other young men were able to begin careers as planters in the same way. For a few years before the Crash, planters' sons utilized these economic institutions as an alternative to the family.[27]

Young men entered into new kinds of social relationships that were primarily commercial and intensely competitive. Josiah Nott was relieved to escape his medical practice at home in South Carolina, where he "knew everybody" and had to spend a lot of time with his patients, some of whom were "particular friends." In Alabama his practice was a "mere money transaction" and therefore "less troublesome." Henry Tayloe, a hard-edged planter who lent money at steep interest rates, cheerfully admitted to his brother, "I have no scruples towards the Alabamians" because "I came here to make money." William A. Lenoir tried to explain these differences to his brother Thomas, who remained at home in Wilkes County, North Carolina. Thomas was unprepared for the "intrigues" of a "business community" because he had been surrounded by kinfolk and friends who tolerated his easygoing ways. William warned his brother that he had to become more "brisk and active" to succeed.[28]

Many men were pleased at first with the Southwest, which seemed to be a veritable "*new world*" brimming with opportunity and possibility. Eli Lide told his brother-in-law that "it would do your very heart good" just to look at the Alabama soil, and Richard C. Ambler declared that his land was "perhaps the very best in the world." Israel Pickens bought one thousand acres of "level rich land" and wanted to buy another "most delightful tract." Great wealth, or economic independence at the very least, seemed to be within reach.[29]

Men experienced the landscape itself in new ways. In contrast to the desolate seaboard, they thought the forests and prairies of the Southwest were "grand and sublime," "wide and romantic," or the "prettiest country . . . I have ever been in," and they found them aesthetically pleasing because of the rich soil. Samuel Townes rode through his fields and pronounced Alabama a "glorious farming country." Some men found the hustle and bustle around them especially stimulating. Josiah Nott declared that there was "something exhilarating in the prosperity and activity of everything about us, when compared to the lifeless despondency of Carolina and the other old states."[30]

Planter men also enjoyed the psychological independence of their new lives, the liberation from the confining ways of the seaboard. A North Carolinian expected to "rise and soar like an eagle" in the Southwest, free from the entanglements of his family, and Israel Pickens wrote that his log cabin was simple but "independently situated." After Samuel Townes arrived in Alabama in 1834, he practiced law in the village of Marion and cultivated a plantation in Perry County. His cotton crops were "*superb*," and he was in excellent spirits, pursuing his law practice with verve. Perhaps most important of all, he had a place of his own at last. "I never felt more independent & cheerful in my life," he announced.[31]

But most settlers, no matter how enthusiastic, had to adjust to the mundane aspects of frontier life. After so many extravagant expectations, some disappointments were inevitable. When slaves began clearing land and planting crops, some masters shared in the physical labor, apparently for the first time in their lives. After

O. G. Murrell worked in the field during a cotton harvest, he told his brother he had never done so much manual labor, and he found it depressing. Richard Archer had to help his inexperienced overseers run his plantation, and he too said that he had never worked harder. Another planter supervised his slaves and worked "a good deal himself," according to his wife, "*splitting rails the most of last week*" to her surprise. But most planter men did not depart so much from their typical roles, which involved supervising slaves through an overseer rather than working with them, and the few who did usually ceased these exertions as soon as plantations began running smoothly.[32]

Other aspects of frontier life, such as the climate, proved to be permanent features of life in the Southwest. John J. Ambler, Jr., grumbled that Alabama was as hot in October as Virginia was in August, and he was right; the Southwest was hotter and more humid than the Piedmont regions of the seaboard. Many men suffered from ill health, the chills and fevers brought on, they believed, by the "broiling sun," "burning rays," and "sickly atmosphere." One disillusioned settler decided that the richest land was also the unhealthiest. The Southwest was in fact more unhealthy than the seaboard: its climate, contaminated water sources, and poor sanitation all contributed to sickliness, and falciparum malaria, the variant prevalent in the region, was more dangerous than the vivax malaria common in the Piedmont areas of the seaboard.[33]

Some planters found that the soil was not as rich as they had expected. Henry Tayloe was so delighted with his new plantation that he named it Adventure, but after two poor harvests he considered returning to Virginia. Thomas Brown discovered after three small harvests that the climate of middle Florida was not suited to the cultivation of sugar cane. The cost of living in the new land took some by surprise, such as Richard Hackett, who found many of his expenses in Shreveport, Louisiana, beyond his means.[34]

Some frustrated migrants began to regret coming to the Southwest. One man acknowledged that there was money to be made, but his expenses, especially doctors' bills, were greater than in the seaboard. He advised his kin in North Carolina not to migrate unless they could come "full handed," that is, with a lot of money. Others attempted to rationalize the decision to migrate. After a bad harvest Willis Lea admitted to his father that his home in North Carolina was preferable in many respects but concluded weakly that human beings were never satisfied no matter where they lived. When signs of disappointment appeared, relatives in the seaboard tried to convince the prodigal to return. As Benjamin Yancey's health faltered in Alabama, his aunt Louisa Cunningham, who said she would not live there for all of the wealth in the Indies, urged him to come home to South Carolina.[35]

Men rarely discussed the opinions of their wives and female kin as they pondered these issues. Those who did mention them usually did so in passing, and they often dismissed women's reactions to the new country. James R. Deupree of Missouri wrote that "all the men is very well pleased but the women is not very well satisfied," but he planned nonetheless to purchase more land in the area. W. F. Hunt related that his sisters were "rather dissatisfied" with Florida but proclaimed in the next breath, "We are all still satisfied." A visitor to Tennessee concluded that his relatives liked their new home, writing, "They and theirs were all quite well. Ed full of life and spirits; Lucy downcast and sober."[36]

Women had cause to be downcast and sober. The Southwest was certainly a new world for women, but it was preeminently a world without kin. Torn from the intricate networks of the seaboard, many women were sad, disoriented, and acutely homesick in the early phases of settlement. Mary Ann Taylor, who arrived in the Southwest in 1834, felt "surrounded by strangers with whom I have not a single congenial feeling," while Louisa Gray, whose thoughts often returned to her native North Carolina, wrote that living in Alabama was more like "fiction" than reality. One young North Carolinian confided to her mother that she felt their separation even more keenly after she and her husband settled in Holmes County, Mississippi. "It seems so unnatural for me to be living so far from you that I can never visit you or have the pleasure of your company at my house." For women these separations were a violation of the ties of nature that bound relatives to each other, not a source of liberation.[37]

Furthermore, women found the landscape dull, empty, even repellent. They repeatedly described the forests and prairies of the Southwest as "dreary," and the countryside gave them a sense of isolation that several called "penitential." A planter's daughter visited her sister's home in the "wild country" of central Alabama and shuddered, "You could see nothing but woods all around." Women disliked the landscape for another reason: it lacked the kinfolk who had given meaning and beauty to the seaboard environment. Mary Drake lamented that the loss of her "large and respectable circle of relations" in North Carolina made Alabama nothing but a "dreary waste."[38]

Like men, women were plagued by bad health in the new country. The climate in Louisiana, according to Adelaide Crain, robbed young women of the "beautiful bloom" in their cheeks. A Mississippian reported that his entire family fell ill but his wife "suffered the most," adding, "She is not yet well." Serious illnesses struck men and women both, but it was a woman's duty to nurse the sick, and that increased the amount of work they performed. Lewis Caperton said that his wife's health had been fairly good until he became ill, when she "broke herself down nursing me."[39]

Women were completely unprepared for the material deprivations of the initial phases of settlement, a topic they discussed at length in their letters. Like women on other American frontiers, their reactions to the Southwest depended to some extent on their previous experiences. Most planter women were accustomed to some level of comfort and "conveniency," although few had lived in real luxury. They were nonetheless more affluent than the farmer's wives who helped settle the Southwest or other frontiers, so the change in their living standards was proportionally greater. And none of the planter women examined here had lived in frontier areas before, as had some white women on other frontiers.[40]

Planter women were shocked by the conditions of their new homes. Most families lived in tents when they first arrived and then moved into log cabins with floors made of hard-packed earth or wooden planks. Many cabins had only a single room, while others consisted of two rooms linked by an open breezeway or "dog trot"; some were crawling with spiders and insects. Men probably did not enjoy living in such places, either, but these living conditions affected women more adversely because they spent more of their time indoors. Mary Boykin Chesnut, whose father, Stephen D. Miller, migrated in 1836 when she was a teenager, recalled that the family's log house "looked as dismal as a prison." Another woman complained of the cold in her log cabin and longed for the "big warm fire" at her father's "fine house" in South Carolina. Mrs. Edward Wells from Virginia burst into tears and made a "fuss" when she first saw the cabin her husband constructed for the family, but, as he put it, she "had to stand it."[41]

These homes were quite a contrast to the residences of the seaboard. One Floridian admitted that his wife and daughters could not get used to living in a cabin and sleeping on the floor. For more affluent planters' wives, the contrast was even more pronounced. Mary Miller, wife of Stephen D. Miller, left an elegant home furnished with marble tables for a cabin where the beds, tables, and chairs were constructed of pine logs. Many women looked forward to the day when they would move into more comfortable homes. Harriet Eggleston ordered expensive tablecloths even though her husband said that "anything would do for a log cabin." She retorted that she did not expect to live in one for the rest of her life.[42]

Within these dank, dirty, and badly lighted structures, women did the kind of manual labor that their grandmothers had done in the eighteenth-century seaboard. One woman acquired the household skills necessary for the frontier life from her sister-in-law, who had already moved to the frontier, but many women must have taught themselves by the age-old method of trial and error. They learned to prepare food over an open fire, ground corn into meal by hand if there were no grain mills nearby, and made clothing for their families rather than purchasing it as they had done at home. One planter's wife rose before dawn every day to begin her chores, which included weaving homespun for the white family and the slaves, as well as raising livestock and poultry.[43]

Slave women assisted plantation mistresses in these chores, but the sex ratios within the slave population were imbalanced, which probably reduced the number of female slaves who worked in the household. The ratios for the slave population between the ages of ten and fifty-five in the censuses of 1830, 1840, 1850, and 1860 for six rural counties in Alabama, Mississippi, and Texas reveal that most counties had substantial male majorities. (See Table 4.) The most imbalanced ratios tended to occur in the first two censuses; Perry County, Alabama, registered the greatest imbalance in 1840, when there were 116 male slaves for every 100 female slaves. The ratios evened out by 1860, when most of the counties had ratios approaching the norm, a slight female majority. Slave men evidently outnumbered slave women among migrants to the Southwest, but other factors may have affected sex ratios: women may have died in greater numbers in rural areas, and planters and slave-traders may have sold female slaves to urban areas, where they outnumbered slave men in selected cities for most of the censuses between 1830 and 1860.[44]

Several comments by whites also indicate that comparatively few slave women in rural areas worked as house servants. In the 1830s a traveler noted that slave women in Mississippi were "generally ignorant of housework" because the first female slaves brought to the region were accustomed to field labor. Furthermore, masters diverted some slave women from household labor to field labor, especially during the first years of settlement. Richard C. Ambler decided to put slave women and girls to work clearing land after he discovered they could do many chores, such as chopping down undergrowth, as well as men. Samuel Townes sent several house slaves into field labor and told his wife Joanna to make all of the slaves' clothing "to keep Sally [a seamstress] in the cotton field." Planter women may not have concurred with these decisions. Jane H. Woodruff, for instance, feared giving birth to a child with "no one with me but an old negro woman out of the field."[45]

Even those planter women with house slaves were not exempted from hard work. Ann Gordon Finley, who moved with her family from North Carolina to Alabama, had house slaves, but she slaughtered and cleaned hogs herself; she also noticed that a neighbor looked "weatherbeaten" even though she too ran her household with the assistance of slaves. The sheer volume of tasks overwhelmed some planter's wives. Ann Archer, wife of planter Richard Archer of Mississippi, had house slaves but admitted that her household duties were sometimes so pressing that she neglected her husband. Alabaman Charles Cooke bragged that his wife could run their household efficiently with only three slaves, as opposed to the six slaves who worked in his cousin's home in Virginia. In this household, all four women—the mistress and the three slaves—probably did more work than they had done at home.[46]

Plantation mistresses had to feed, clothe, and nurse slaves, chores that set them apart from other white women. Slave men and slave women did the hardest work on the frontier, of course, and they spared plantation mistresses many of the strenuous, exhausting, and dangerous chores that other white women had to perform. For example, the family correspondence examined for this study contains no account of a planter woman felling trees, a chore other white women performed on American frontiers. It is difficult to judge whether planter's wives worked harder than other white women in the Southwest (whose lives we know virtually nothing about) or white women on other frontiers, but they certainly did different kinds of work.[47]

Planter women supervised the feeding of slaves, and they sometimes prepared meals themselves. Jane H. Woodruff, the sheltered woman from Charleston, fixed meals for at least a hundred slaves on a plantation in northern Florida. She fed them herself and shared their diet, even when the meat was infested with maggots. Many planters' wives delegated weaving cloth to their slaves, but some worked at the looms themselves, and most mistresses cut, dyed, and sewed slave clothing. They repeated these chores annually long after the frontier phase ended, distributing wardrobes every spring and fall, and they nursed slaves if they were sick or injured. When the slave Mary fractured her arm, it was her mistress, Eliza Townes Blassingame, who remained at home to care for her, rather than the master or the overseer. Once again, women whose families owned many slaves probably worked the hardest of all. Jane Woodruff nursed the family's many slaves even when she was in the advanced stages of pregnancy. When more than half of them were downed by fever, she cared for them herself, her time "entirely taken up" with caring for the sick. Thoughtful men sometimes noticed the toll that these chores exacted from planter women. One Alabama planter observed that they soon became nothing but "domestic industrious housewives." "How can it be otherwise," he asked, "when labor alone is the only employment to cheer up life in this dreary solitude?"[48]

These experiences probably help explain why planter women were more sensitive than planter men to the adjustments that slaves had to make in the new country. Both sexes had daily contact with slaves in the pioneer phase, but women more often than men acknowledged the fact that slaves became ill. Ann Finley described not only her husband's sickness but also that of two of their slaves, one of whom had to learn new kinds of work at age forty. Samuel Townes, however, fell ill with fever but insisted that his slaves were pretending to be sick so they would not have to work. When seaboard relatives visited the frontier, these differences between the sexes were also evident. Caroline Gordon, the sister of Ann Finley, relayed that the Finley slaves thought Alabama "the most doleful place to be found on earth." William Townes, one of Samuel's brothers, visited Alabama and decided that the slaves were "fat and saucy and all satisfied to stay here," with only one exception, Jim, who "talks about home as usual."[49]

The discrepancies between the sexes are especially striking in the correspondence of the Brown family of North Carolina, who moved with their slaves to Maury County, Tennessee, in 1841. Mrs. Allen Brown went reluctantly to the frontier and told her sister that she had returned "to my native hill" in her dreams many times. She could "scarcely realize" that she was "four hundred and fifty long miles from you." While her husband insisted not only that everyone was "well pleased" with Tennessee but that there was "no reason" anyone could be displeased, Mrs. Brown knew that she had "left too many of my enjoyments behind to be satisfied soon." After the entourage was settled in its new home, Mr. Brown reported that his slaves had decided that "the country is not so bad as they expected," but his wife sketched a different picture. She was relieved that a pregnant slave had withstood the trip and given birth to a healthy daughter, and she acknowledged broken family ties by sending a message to the woman's husband and reassuring several other parents that their children were safe. Like many plantation mistresses, she sent greetings to white relatives to pass on to slaves, writing in one letter, "All the negroes send howdy to you all white and black." After several more such messages, she concluded that her missive resembled one that a slave might write, "all love and compliments," and then made this abrupt disclaimer, "You must excuse it."[50]

It is difficult to interpret these last oblique remarks, in which Mrs. Brown almost, but not quite, compared herself to a slave. Did she want her reader to excuse this analogy, or the number of messages she was sending to slaves, or perhaps the effusiveness of the messages? It is impossible to say. The remaining correspondence reveals nothing about Mrs. Brown's racial attitudes; it does not even tell us her first name. But she suggested that she, like her slaves, missed her absent relatives very much, and she implied that migration affected the planter family and the slave family in similar ways, separating kinfolk who loved each other. She did not express the outrage of planters' sons from the seaboard who compared themselves to slaves, and she backed away from the further implications of the analogy—that the ambitions of planter men sundered the white family as well as the slave family and that planter women, like slaves, could not prevent it.

The poor communication between the sexes that marked every stage of migration recurred during the early years of settlement. Thomas Meade wrote his sister that his wife, Anna, had adjusted to their plantation in Mississippi, but she added in a postscript to the same letter that she was "quite lonely." Men certainly pressured women to accept the deprivations of the frontier phase. Another planter's son scolded his sister for being "such a poor soldier in the strifes and vexations of a settler's life." Some allowed women a period of time to compose themselves. Thomas Tabb knew that his mother was discontented in Alabama but thought that she would like it better once "her mind becomes settled." But almost all men expected that women would eventually accept the situation, and the sooner they did so, the better. George Townes of South Carolina believed that his sister Eliza Townes Blassingame "fortunately and wisely" was "reconciling her mind" to her Alabama home in early 1834, and several days later brother Samuel, who was still in Carolina, thought that "she could have no motive for disguising" her feelings.[51]

Planter women did have reason to mask their feelings, of course. Caroline Gordon of North Carolina could tell that her sister Ann Finley missed home very much but that she tried to conceal her feelings from her husband. When Mrs. Finley came home from Alabama with her children for a long visit, they "seemed to dislike the idea of going back so much that they delayed it as long as possible," and one of the children wished that "he was a partridge that he might fly back to see us," but Ann Finley and her children did go back. Most women tried to accept the situation because they had no other choice. Sarah Lide Fountain strove to be "resigned" to her new life in Alabama and hoped that "it will turn out for good in the end."[52]

The Southwest was also a new world for slaves, a world of deprivation and danger. Their living conditions during the pioneer phase were even more difficult than those of their masters and mistresses. They lived in dingy cabins, ate irregularly, worked long hours clearing land, plowing, planting, hoeing, and picking crops, and through it all they mourned for their relatives and friends in the seaboard. Slaves were at the mercy of their owners, depending on their responsibility, good sense, and advance planning to provide for them, and masters were not always reliable. Joseph Woodruff, a Florida planter, did not bring enough food from South Carolina, so his slaves had to fend for themselves for almost a week.[53]

Slaves may have been more vulnerable than whites to the perils of living on the Southwestern frontier, as illustrated by the fate of twenty-two slaves owned by Richard and John Ambler. The two brothers left Virginia in 1835 for land in Perry and Dallas counties, Alabama, they bought sight unseen for $6,120. Each of them provided eleven slaves for the workforce, but Richard did most of the work, taking the slaves to Alabama himself, and he was especially eager to establish his "independence." The Ambler slaves might seem more fortunate than most because both masters had studied medicine, and Richard Ambler was a practicing physician. But soon after he arrived with the bondsmen most of the slaves contracted fever, and several died despite his treatments, while Indians who lived in the forests at the edge of the plantation murdered several other slaves. When John Ambler appeared several months later, at least eight slaves were dead, and he found the rest in despair, fearing that they had been "abandoned to disease and death."[54]

Further Migration

As their frustrations mounted, many planter men began to consider either returning home or moving farther west. Some wished to go back but could not afford to do so, such as William S. Brown, originally of Virginia, who was resigned to living in Missouri for the rest of his life. Others decided to give up the venture and go home. As one man commented, they were "coming back after cooling."[55]

As might be expected, many returned because they did not find the prosperity they sought. Some met disaster soon after they migrated, such as the Ambler brothers, who spent little more than two years in Alabama before they sold their land at a great loss and returned with their remaining slaves to Virginia. Others struggled for a long time before giving up. Andrew P. Calhoun lived on his Alabama plantation for over twenty years, accumulating more and more debts, before he straggled back to South Carolina on the eve of the Civil War. Yet others went back because safer opportunities beckoned from home, such as the North Carolinian who returned from Tennessee to go into business with a kinsman.[56]

Sometimes the seaboard became more attractive as men grew older and had families of their own. Eustace C. Moncure, Jr., lost two children in Texas and was having trouble making ends meet. To his father he admitted that he was at his wit's end and was considering his brother's proposal to come back to Virginia to farm. After Benjamin C. Yancey became seriously ill in Alabama in 1840, he relocated in his native South Carolina. Ten years later he came back to Alabama, but he moved to Atlanta in 1856 to find better schools for his children. Yancey was then thirty-nine years old; more than twenty years had passed since he first left the seaboard for the Southwest.[57]

Planter men who returned to the east sometimes met with further disappointment as their goals continued to elude them. John Ambler expected his financial troubles to haunt him into "remote old age," and, as he feared, he eventually lost every one of his slaves and was swamped with debts until he died in 1854. Lewis Caperton almost died of typhoid fever in Texas and went back to Virginia broken in health and spirit. Another Virginian came home a "disappointed and unhappy man" and lapsed into a state of dependency, living off his family. It is beyond the scope of this book to trace the destinies of these men, but they seem to have blended once again into the collective families of the seaboard.[58]

Planter women were enthusiastic at, if not elated by, the prospect of going home. They sometimes spoke up when their husbands or male relatives began to discuss the idea. One matron let it be known that she was "very anxious" to return to North Carolina when her kinfolk expressed dissatisfaction with Tennessee. After Mrs. Lewis Caperton went home with her family, she did not hide her memories of illness and hardship. Her husband related that she "abuses Texas whenever she speaks of it." In a few instances, women made these choices themselves. When Mary Miller, the mother of Mary Boykin Chesnut, was widowed in the Southwest, she moved her children and slaves back to South Carolina to live near her relatives. Women were only too ready to go back to the embrace of the family.[59]

Slaves rejoiced at the prospect of returning to the seaboard. When Matthew Wallace and Ann Meng Wallace of Union County, South Carolina, decided to move to Mississippi in the early 1850s, their slaves dreaded the trip. During the four-week journey, the party had such bad luck—broken wagon wheels, injured horses, attacks by wild animals—that the slaves decided the venture was jinxed. Misfortune followed the party to DeSoto County, where Matthew Wallace died of pneumonia. Ann Wallace then announced to the slaves that she had decided to take them all home, whereupon they started "dancin' and a hollerin'" and packed the household goods with dispatch. Many years later, Sallie L. Keenan, the daughter of one of these slaves, remembered her mother's great relief at coming back to the east.[60]

Rather than going home, some dissatisfied white men decided to keep moving to improve their lot. One western planter watched many of his neighbors rushing to Texas in 1840, their "gold dreams . . . now realizing." Those who migrated within the Southwest headed to the north, south, and west, sometimes moving for short distances before making another long-distance move, but the general movement was farther west, toward the unspoiled lands of Texas. The motives of men had not changed: they wanted personal independence, and they wanted to get rich. Planters searched for fertile land in healthy locations; those who also practiced law looked for the "place of places" in towns throughout the region, while physicians made discreet inquiries about locations where the population was wealthy and the climate harmful.[61]

These moves resembled the initial departure from the seaboard in most other respects. Men consulted their male relatives, friends, and colleagues, found information in newspapers or periodicals, or gathered it first-hand during exploratory trips.[62] Planters had not learned much about agriculture from their experiences in the Southwest, for they were as impatient with its soil as they had been with the fields of the seaboard years before. When John Horry Dent left South Carolina for Alabama in 1837 in search of independence, he expected that he would not have to move again, but he relocated within Alabama at least once to a more profitable plantation, and by the 1850s, when Dent was in his forties, he was ready to make a major change. He complained that his cotton and food crops were abominable and that attempts to fertilize such "precarious lands" were useless, so he decided to examine land in Mississippi and Texas.[63]

Planter women opposed further migration for several reasons. Those who had the good fortune to live near relatives were grieved to have to say goodbye yet again. After Jane Boyd watched her relatives move west from Alabama, she lamented that too many people forgot the real aims of life, such as helping others, and sought only "wealth and honors," and she continued to believe that it was "natural" for kinfolk to live near each other. Other women had developed close ties with some of their neighbors, and they found them hard to forsake. One woman thought that her father's decision to move from Tennessee to Mississippi was "a great mistake" because of the lack of pleasant neighbors there. Still other women objected to mobility itself, to the upheaval of uprooting a family once more. After nine years in Alabama, the women in the Lide family were furious when their male kin wanted to move to the Red River Valley in Arkansas. One of the Lide women declared that she expected to spend the rest of her days "moving about from place to place."[64]

As these examples suggest, the sexes continued to hold different views of geographic distance. Whereas women were dismayed at the prospect of moving several dozen miles farther west, deeper into the wilderness, men accepted even greater separations from the seaboard as necessary and good. Samuel Townes, for instance, considered moving from Alabama to Louisiana to make more money, remarking that it was only "a few hundred more miles" from his relatives in South Carolina. Anne Dent, however, charged her son John Horry not to move from Alabama to Texas because his children would "grow up perfectly rough and uncultivated" and the family would be even more isolated from relatives. Mary Drake, whose father settled in Alabama, criticized her brother for moving several hundred miles away within the state. She felt he had abandoned their father, who was growing senile, and concluded that her brother "has moved so often he cannot be contented anywhere long."[65]

Men excluded women from decision-making in this round, too. Sarah Gayle watched nervously as her husband and brother-in-law discussed moving from Greensborough, Alabama, observing that "they appear to be near a crisis," but her husband apparently did not discuss the subject with her. Women recognized that only other men could influence their male kinfolk. Virginia Gordon knew that only her brother-in-law could dissuade her brother from moving from Mississippi to Louisiana. In one unusual case, a man gave his teenaged daughter the choice of remaining in Tennessee with him or accompanying her brother, sister, and brother-in-law to Arkansas in 1860, but there is no record of her response.[66]

Further migration within the Southwest resembled the initial departure from the seaboard in one other respect: it destroyed family ties within the slave population. Planter men usually ignored the ties of blood and marriage among bondsmen, separating spouses, siblings, parents, and children as they went on to other frontiers. Planters who remained in one place also paid little attention to family bonds when they trafficked in slaves. Henry Tayloe of Alabama, who participated actively in the slave trade within the Southwest in the 1830s, bought and sold many slaves with a stony disregard for family ties. As Samuel Townes prepared to sell the slave Daniel, whom he owned jointly with his brother George F. Townes, he thought the transaction should be "governed entirely" by economic interests.[67]

Once again, plantation mistresses were more aware of how these events affected slaves. Marianne Gaillard's husband, Thomas, was in serious financial trouble by 1844 and had to sell many slaves from his several Mississippi plantations. She thought it was "a great affliction to see our Negroes leaving us to be sold; I cannot part with them with as little concern as though they were cattle," implying, perhaps, that her husband felt little concern for them. She also feared that one of her husband's "ruthless" creditors would buy up more slaves, separating "the Mother from her Sucking Child & the Husband from the Wife." Unfortunately, Mrs. Gaillard seems to have had little influence on her husband's fiscal decisions; she implored her male relatives to come out from South Carolina and assist him.[68]

Even as planter men and women pressed on with the daily business of running plantations and households, they tried to give shape to family life in the new country. The sexes had different visions of what the family should be, however, and they met this challenge with inequitable resources: men still had the decision-making power in the family. But men too discovered that unexpected outcomes awaited them.

4

A Little More of This World's Goods: Family, Kinship, and Economics

Planter men who remained in the Southwest wished to break free of the family, while planter women tried mightily to preserve familial ties, but the material and social conditions of their lives made it impossible for either gender to achieve their separate goals. Most men discovered that it was difficult to prosper without the family's economic resources, while most women could not re-create kinship networks because of the dispersal of the family across geographic space. Family life broke down and was reconstituted; some elements remained more or less the same, but most elements changed. The pact between the sexes that had prevailed in the seaboard was no longer in place.

The Structure of the Family

One element that changed was the long-term residence patterns of the household. Relatives shared quarters in the early months of settlement, as did the three first cousins and their second cousin who lived together in Texas while they decided where to settle permanently, or the Williams family from South Carolina who lived with cousins during their first winter in Florida while their own house was being constructed. These living arrangements were only temporary, however, a response to the housing shortage common on many frontiers.[1]

As the decades passed, nuclear households became the norm in the Southwest. Federal census returns for 290 planter households in six Southwestern counties indicate that nuclear households were the most common form of household structure for the censuses of 1840, 1850, and 1860. (See Table 5.) These 154 nuclear households constituted slightly over half (53 percent) of the sample. Complex households, sixty-five in number, made up less than one-quarter (23 percent) of the sample. Seventy-one households were ambiguous in structure—neither nuclear nor complex—and constituted 24 percent of the total. (The percentage of ambiguous households dropped from the seaboard percentage because the censuses of 1850 and 1860 provide more information than earlier censuses.) The distribution of household types remained roughly the same throughout the period. The range in household size, from two to fifteen persons, was smaller than in the seaboard (two to twenty-three persons). The average household size, six persons, was also smaller than the average of seven persons in the seaboard, but it was typical of other white households in the Southwest.[2]

It may well be that nuclear families are the norm on frontiers, or that they are a function of the comparative youth of male heads of households, but the values of planters' sons who wanted to escape familial obligations also played an important part in this transformation. As these households took shape, men and women often took a different view of the issue. Sarah Gayle of Greensborough, Alabama, wanted to raise an orphaned niece to "relieve my heart of a load of tenderness, of gratitude, of duty" to her deceased mother, who "would have cherished this daughter of my unfortunate uncle." The young woman was in poor health but worked at some sort of "trade," probably as a seamstress, and she was willing to support herself. John Gayle, Sarah Gayle's husband, nonetheless refused to allow the woman to join the household.[3]

Significant changes also occurred in relations *between* households as individual families became more isolated than in the seaboard. The kinship networks of the Southwest were imperfect and incomplete: the range and number of available kinfolk were never as great as at home. Women tried to raise their children to forge attachments to relatives, as did the Louisianan who encouraged her children to "love everybody kin to them," but there were not many kinfolk nearby to love. In contrast to seaboard families who lived surrounded by kinfolk, Southwestern families lived near a few relatives or none at all. The planter family was reduced to its nuclear core.[4]

Women felt this transformation most keenly, of course, and wrote about it frequently. Their social networks, built upon kinship, had virtually collapsed. Many felt lonely in the Southwest, even in the company of their own families. Fanny Polk was "depressed" with no one to talk to except her husband on the Tennessee frontier, and she did not have "the courage to venture out" and meet her neighbors. Mary Rives, whose family had settled in Hinds County, Mississippi, bemoaned the fact that she was "alone, except for my own immediate family." Women's sense of isolation continued for many years after settlement. A North Carolinian whose family had lived in Mississippi for fifteen years lamented that she was "alone, far from relatives except her own children."[5]

Many planter women missed the comparatively easy access to kinfolk they had enjoyed in the seaboard. Anna Meade spent the first three months of her married life alone with her husband on a Mississippi plantation, whereas back in Virginia a week never passed without a visit from her relations. She enjoyed a visit from her nephew but admitted that it was no substitute for the full complement of kinfolk she had left behind. Alabaman Ann Finley envied her sister in North Carolina because she could spend hours with "dear relations so near" whenever she pleased. Mary Drake, also of Alabama, missed the spontaneity of visiting customs of the seaboard. She longed for the days when one of her children would suddenly run into the house shouting that aunts, uncles, or cousins were arriving.[6]

Women especially missed the personal, face-to-face contact with their female kin. Ann Finley wanted some of her relatives nearby so she could "seat myself and have a good talk" with them, and she told her sister that she knew of no enjoyment greater than "a sweet intercourse with yourself and family face to face." Correspondence, as a substitute for conversation, became an even more important link between women than it had been in the seaboard. Receiving letters from relatives was Mary Rives's "greatest worldly pleasure," and Mary Drake said that letters from her kinfolk in

North Carolina, especially her dear sister, were "the greatest comforts I have left." Adelaide Crain, who reread her cousin's letters until she could "almost repeat them verbatim," thought that she would "lose my identity" if she could not write open, expressive letters in reply.[7]

The fear of being forgotten by relatives ran like a dark undercurrent through the correspondence of many women in the Southwest. So far from their homes, with their kinfolk scattered across the South, they felt that their families were breaking apart. When one woman sat down to write to her niece in North Carolina, she discovered that "it brings you all so clear before me that I cannot keep from crying." All of this suggests that a certain sadness and yearning marked relationships within the planter family in both its Southwestern and seaboard branches.[8]

Visits between Households

Visiting, one of the most important links between families, became more difficult in the Southwest. The region lacked good roads and extensive railroad lines for much of the antebellum period, and the terrain seemed forbidding to many settlers. Marianne Gaillard once said that a distance of twenty-five miles in Mississippi seemed much longer than the same distance in her native South Carolina. To many women, geographic space seemed unmanageable, unbridgeable, as it had never been before.[9]

The problem was compounded because planters tended to settle on tracts of land hundreds and even thousands of acres in size, so that many households were farther apart than they had been in the seaboard. John W. DuBose, who grew up in South Carolina surrounded by relatives, said that his family's "'near' neighbors" lived at least a mile from their new plantation (which contained 890 acres) in Marengo County, Alabama. Although the figures vary, the average plantation in the Southwest was probably larger than its counterpart in the seaboard, and many plantations grew even larger in the late antebellum era. The average size of farms owned by slaveholders in Hinds County, Mississippi, increased from 204 acres to 318 acres between 1850 and 1860, and the average size of plantations in 1860 in four counties in Alabama's Black Belt ranged from 651 to 770 acres. Very large plantations also seem to have been more prevalent: in 1860, for example, Marengo County, Alabama, contained sixty-two farms at least 1,000 acres in size, while Charlotte County, Virginia, contained only twelve.[10]

The architecture of homes also made visiting difficult, because the log cabins many planters lived in could accommodate few guests. Samuel Townes lived in a two-room cabin, and he suggested that his mother stay with his sister and brother-in-law, who had four rooms, when she traveled to the Black Belt in 1835. Another Alabama matron could not house guests comfortably in her one-story residence. Even in the late antebellum era some planters lived in cramped houses that made visits awkward. Harriet B. Eggleston of Mississippi postponed a visit from her sister-in-law in 1859 because she could not accommodate her. In these homes there was literally and figuratively no room for the congregations of kinfolk who had filled the houses of the seaboard.[11]

Furthermore, women were even more immobile than they had been at home. They still could not risk social disapproval by traveling alone—which was an obstacle in and of itself—and they faced new physical dangers on the frontier. Creek Indians attacked stage coaches in Alabama and murdered the passengers, leaving the corpses to rot by the roadside. In Mississippi vagrant whites sometimes attacked women in carriages. The countryside was inhabited by wild animals, and the tropical lushness of the landscape itself could be menacing. The canebrake in the Black Belt was ten feet high, and it was so thick that men on horseback got lost in it as they moved from one plantation to the next.[12]

So women traveled in the company of men, even for short distances. One young woman was accompanied by a male cousin during a twenty-mile journey from Shreveport, Louisiana, in the 1840s. Nor could female children leave the plantation without a protector: a planter's daughter walked with one of the family's favorite male slaves on her way to and from the local schoolhouse, which was several miles from her home. A pathetic comment on female dependency and immobility came from seven-year-old Eliza Irion when she could not visit an aunt who lived fifteen miles from her Tennessee home. She missed her aunt so much that she wished that she were an animal—a cow, to be specific—so that she "could go all by myself."[13]

Even in extraordinary circumstances, women's mobility was still dependent on the approval of men. Mary Ann Whitfield, wife of planter Gaius Whitfield of Marengo County, Alabama, spent many evenings alone in their mansion while her husband tended to his various properties. One spring night in 1835 a drunken white man stumbled uninvited into the house; a frightened Mrs. Whitfield fled on foot to the home of a neighbor on the next plantation. Evidently it was highly unusual for Mrs. Whitfield to go anywhere without her husband's permission. Even though she thought she was in danger, she asked his forgiveness for "going without your knowledge."[14]

Much as planter men decided who would join the household on a permanent basis, they decided who would visit the household and whom their female kin would visit. Visiting, according to Alabaman Hardy Wooten, was a foolish waste of time, and it was not a priority for men who had left the seaboard to escape frequent interactions with the family. If anything symbolized the breakdown of the pact between men and women, it was this unwillingness to facilitate visits. Widow Mary Drake, who lived with a kinsman, discovered that he would not permit her to invite many guests into his home, much to her chagrin, and William Blassingame, Samuel Townes's brother-in-law, sometimes refused to let his wife, Eliza, visit Samuel, who lived in the same county. Nor did men always cooperate with women in arranging visits across long distances. Sarah Gayle, who grew weary of her husband's "endless wanderings" on political and legal business, passed "seven long years" without visiting relatives who lived some two hundred miles from her home in Greensborough, Alabama, and Mary Ann Black of Tennessee reported sorrowfully that she had not seen her sister, who lived three hundred miles away in Mississippi, in seven years.[15]

Women therefore savored visits all the more when they did take place. Sarah Gayle fondly recalled a rare visit with one of her aunts, who was "so gentle, indulgent and kind" to her and her children; "Would to Heaven I had some way of going to see them," she exclaimed. She also missed her sister-in-law "in the thousand every-day occurrences which make up more than half one's life," adding, "I want her sympathy, her ready attention." Ann Finley was starved for the company of her female kinfolk, and the intensity of her loneliness startled her relatives when they visited her in 1859. Her sister Caroline Gordon observed that Finley wanted her guests to "devote ourselves to her"; "she is so fond of talking and grumbles if we even take time to write letters." Mrs. Finley was deeply disappointed when other visits failed to materialize. When she learned that two male relatives were traveling alone from North Carolina, she cried, "Can it be that none of our female relations will come out with them?"[16]

In the Southwest men ranged widely from home, walking or riding at will just as they had in the seaboard. The contrast in the mobility of the sexes is especially evident in the marriage of planter Benjamin Whitfield, a North Carolinian and cousin of Gaius Whitfield who settled in Hinds County, Mississippi, and his wife, Lucy Hatch Whitfield. The owner of 140 slaves, Mr. Whitfield was active in various state and local associations, such as the Baptist State Convention, and he served as a trustee of Mississippi College, so he journeyed frequently from home on business matters. Mrs. Whitfield spent most of her time at home, rising before sunrise to begin doing her household chores, supervising her house slaves, and caring for the twelve Whitfield children. She was so confined to the plantation that she onced passed an entire year on the place without seeing another white woman.[17]

Many of the material obstacles to visits changed over time, of course, as transportation lines improved, whites forced Indian tribes to cede land and move farther west, and the tide of white settlement pushed back the undergrowth and killed off much of the wildlife. By the late antebellum era, planters in well-settled areas of the Black Belt and the Mississippi Delta could travel almost as easily as their seaboard relatives had a generation earlier, and in these places some women could visit as easily as in the seaboard. In Cahaba, Alabama, a village in the heart of the Black Belt, women could walk to the homes of their kinfolk or friends in the company of a slave or another white woman. They could also make ceremonial visits as well as the lengthy visits they had enjoyed at home, incorporating relatives into their daily activities. But most women were not able to duplicate seaboard visiting customs, and they all needed male approval for most of the visits that they were able to make. The circular, rhythmic mobility they had enjoyed in the seaboard was gone. Most planter women were marooned in the new country.[18]

As geographer Yi-Fu Tuan remarks, it takes time to know a place well, but even the passage of time does not ensure that individuals will develop a strong sense of place, or, I would suggest, affection for a particular place. One North Carolinian whose husband settled the family in Mississippi put it succinctly: "This place still appears strange, and altogether unlike home to me." Here there were no familiar, beloved landmarks, and the homes of kinfolk were few and far between. Geographic space in the Old Southwest was not organized to meet women's social needs. Each population uses geographic space in distinctive ways, and the sprawling plantations of the Southwest reflected the power of their male proprietors and served their social needs.[19]

Kinship, Reciprocity, and Economic Independence

The banking system of the Southwest provided an alternative to the family as a financial institution for only a few years, and the "flush times" collapsed with one of the worst depressions in the nineteenth century, the Panic of 1837. Historians are still debating the causes, which were extraordinarily complex, but the Panic seems to have been the result of movements in the international specie market and a rise in interest rates in England; President Andrew Jackson's fiscal policies had a further disruptive impact. His Specie Circular of July 1836, which required that public lands be paid for in specie rather than in paper money, set off a scramble for "hard money" by thousands of farmers, planters, businessmen, and speculators who had hitherto dealt in paper money. Many banks had been speculating in western lands, and they too searched desperately for specie. Land sales in the Southwest declined almost immediately. In a separate, unrelated development, English bankers contracted credit later that year because they believed too much specie was leaving the country for the United States, which in turn increased pressure on major American banks. In early 1837 cotton prices dropped as a result of disrupted trade relationships between English bankers and American planters. By the spring Americans began losing confidence in the entire banking system, and fearful depositors all over the United States demanded that banks redeem paper notes for specie. Some banks failed, and in May others suspended payment, refused to renew existing loans, and called in loans to bolster their reserves. Borrowers everywhere were caught, and a wave of bankruptcies swept the country.

The depression hit Southwestern planters very hard, maybe harder than other Americans, because so many men had bought property on credit or with bank loans. As banks went under, planters went under, and they took their clients with them. The spring of 1837 was a time of "great distress," according to Needham Whitfield. Henry Tayloe had to sell a number of his slaves and feared that he would be forced to sell one of his plantations "at any price"; he heard rumors that bankruptcies in Mobile were rampant. Cotton prices plummeted from a high of fifteen cents a pound in the 1836 crop year to seven cents a pound in the 1839 crop year; they did not reach ten cents until 1849, and it was not until the mid-1850s that prices held at the levels of the mid-1830s. The Panic also had disastrous long-term effects on the Southwestern banking system, which proved to be particularly vulnerable. Alabama's state banks did not recover until the 1850s, and Mississippi did not even have a functioning banking system for the rest of the antebellum era. This hair-raising series of events eliminated most alternatives to the family as an economic institution.[20]

Many men eventually recognized that the family was still a potent source of capital, assistance, and information and that they had to compromise their ideals of independence. A North Carolinian who opened a law practice in the Southwest found by the early 1840s that that he could not succeed without the aid of "influential" friends or relatives. Another believed that he had not made his fortune in the Mississippi River Valley because of the malevolence of others; even worse, he could find no patron to help him over the rough patches. Samuel Townes of Alabama had to turn back to his family for help, which virtually destroyed his self-respect as well as his dreams of economic independence.[21]

When Townes came to Perry County in 1834, he brought three slaves with him and then purchased others with loans until he owned between fifteen and twenty slaves. He also bought a plantation on the fertile banks of the Cahaba River. After the Crash in 1837 he had to sell some of his slaves, but in 1840 he still had ten slaves left. Over the next several years, he began to borrow money from his brothers and another kinsman in South Carolina, and he rented out some of their slaves in Alabama for a fee—which led to several family disputes—but his finances continued to deteriorate. In 1844 his land was sold at auction, but he told his older brother Henry that he would "endeavor to bear myself like a man—struggle and live on."

That year Townes published a brief history of the town of Marion, which he hoped might bring in some extra revenue. It did not make much money, but it does illuminate his attitudes toward family, wealth, and achievement. He still derided the "slow and painful accumulation of property" which a man had to undergo in a "settled" country, presumably South Carolina. He jeered at the idea that a man's "blood" or his "dead ancestors" mattered in assessing his social standing, and he mocked the typical "scurvy representative" of an eminent old family who tried to live off a name alone. "What has he done—what can he do?"—this was how Samuel Townes and many other frontiersmen took his measure. He recounted the biographies of several neighbors who had risen from obscure origins to become rich planters by dint of "honest industry." (Townes did not say that the industry of slaves and white women helped create the wealth.) Unlike residents of "an old country," these planters could take pride in their own accomplishments.

These attitudes must have made Townes's swift decline over the next several years all the more galling. In 1845 he tried and failed to negotiate another loan from a bank in Tuscaloosa. In desperation he asked friends in the Democratic party for an appointment as a consul overseas, but it never came through, despite the influence of his kinsmen John C. Calhoun and Francis W. Pickens. Townes told his brother Henry that it would be "shameful, nay, disgraceful" to accept any more assistance from his family, that he would rather "black boots and clean out stables."

Then Townes's health unaccountably failed, and he began losing his eyesight. By 1847 he turned once again to Henry, telling him that he was nearly blind and at the edge of destitution. He asked Henry in the name of "kindredship" for help, crying, "I am utterly powerless," little more than a "beggar," and filled with "mortification" at his plight. His younger brother John blamed him for his troubles, coolly remarking that Samuel did not practice the "patience & perseverance" common in "old South Carolina," "the 'land of steady habits.'" Henry, however, believed that Samuel and Joanna Townes "held out as long as possible" before asking the family for help, and he readily came to his brother's aid, lending Samuel money and calling upon his siblings to be generous with their "prodigal" brother.[22]

As other migrants groped toward reestablishing these economic relationships, they discovered that some kinfolk balked, like John Townes, but that other relatives were willing to lend them the money or property they needed. Just as in the seaboard, men sometimes charged interest on loans—most often the rate of 5 or 6 percent that prevailed before the Panic of 1837—but this was lower than the rates of up to 15 percent charged by some banks. Furthermore, relatives usually contracted lengthy schedules of repayment, ranging up to ten years, or simply asked their kinsmen to liquidate debts whenever they could do so. These terms were more generous than any bank could offer, and they could also work to the advantage of the lender: when he needed money, he could call on his kinsmen to repay part or all of their loans on the spot.[23]

The geographic distances between relatives within the Southwest did little to inhibit this exchange of favors, because men transacted most of their business through the mail. Men asked for and received many kinds of assistance from a variety of kinfolk on the frontier. They exchanged gifts and loans of money, slaves, land, and livestock, and they gave and received help from brothers, in-laws, uncles, nephews, and cousins.[24] Many men also began to reestablish ties with kinfolk in the seaboard. Men turned to a host of their male kin at home, and they performed a variety of favors for each other. Southwestern planters hired out slaves locally who belonged to relatives in the seaboard, and they called on kinfolk to help them in matters concerning inheritance. Seaboard relatives provided documentation, such as copies of wills or powers of attorney, for their western kin, and they arranged for property to be transferred to relatives there.[25]

These exchanges did not always go smoothly, of course. Mississippian Willis Lea advised his brother to be careful while dealing with an uncle and protect his own good name. Misunderstandings occurred even among relatives who had done business for years, such as the Burwells of Virginia and Alabama, who became embroiled in an ugly dispute over slaves. Men strove to contain these disputes, however, to protect useful relationships; they sometimes called upon their rediscovered sense of familial obligation. When a disagreement broke out between a North Carolinian and his cousins in Mississippi, a kinsman hoped that it would produce no distrust between them after "our long and early attachment and profound acquaintance."[26]

Just as in the seaboard, these systems of economic exchange trapped men in dependent roles. Virginian O. G. Murrell migrated to Tennessee in the 1830s, but he fell into debt and moved to Mississippi in the late 1840s. There he purchased a plantation with the help of his brother, John, who had remained at home, and he began to turn to John frequently for money and advice. By the summer of 1849 O. G. Murrell was beset by troubles: heavy rains and an infestation of worms were ruining his cotton crop, and many of his slaves were too ill to work. He admitted to his brother that he still did not know how to practice "economy" and keep his expenses down. Finally he relinquished all decision-making power to John: "It is for you to determine what course is best to pursue. . . . I will endeavor to carry out your instructions." As these kinds of relationships evolved, some men abused their power, just as they had in the seaboard. Alabaman Abraham Rencher excluded his younger brother Daniel from his deliberations about a joint business venture, so much so that Daniel complained that he was as ignorant of his brother's plans "as any one of his negroes." Despite their original goals, many migrants ended up re-creating the economic systems of the seaboard family with all of its dependencies and tensions.[27]

The geographic distances separating kinfolk did inhibit the exchange of favors among female relatives, however, because these favors involved sustained labor and personal contact. Women still exchanged small gifts through the mail, but they could not do other, more important favors that had been common in the seaboard. Child-rearing, for instance, was not a single transaction, like many favors that men did for each other, but required the continual participation of female kin. Adelaide Crain, a lawyer's wife in Shreveport, Louisiana, was overwhelmed by "attending to the wants of my numerous family" by herself, and her cousin, Ann Finley, felt worn down, even "crazed," by the duties of rearing her children alone. "I cannot cast my eye in any direction about my house," Finley exclaimed, "but what I see something just waiting for and loudly demanding my attention." Women may have missed their female relatives most acutely when they lost a child. An Alabaman had to grieve alone for her child because she was "entirely unconnected," far from her kinfolk.[28]

Those women who were fortunate enough to have relatives living nearby re-created the systems of exchange they had known at home. Kinfolk helped each other with housekeeping, rendering "mutual comfort and mutual aid." They sewed and quilted together, enjoying conversation as they worked. Their attendance was vital at childbirth because, as Sarah Gayle believed, only other women could fully sympathize with the anxieties of a pregnant woman. Women also wanted their relatives present when serious illnesses struck. As Mary Drake told her sister, nothing cheered a sick room like "the company of an affectionate female relation."[29]

As kinship networks developed in some settlements over several decades, the growing number of available relatives lightened the workloads and the hearts of many women. When Martha Lenoir Pickens left Fort Defiance, her home in North Carolina, and settled with her husband, Israel Pickens, and their children in the Alabama Black Belt in the late 1810s, she acutely missed her female relatives at home. She longed to help raise a niece who was her namesake and urged her sister to visit her, but Pickens saw little of her kinfolk before she died of fever in 1823. By the time her daughter Julia Pickens Howe raised her own children in the early 1840s, however, other Lenoir and Pickens relatives had settled in the Black Belt, and they helped rear her offspring. As one of her cousins told her, "it is a good thing to have friends and relations about you" when a woman was raising small children. But this situation proved to be temporary: the family moved to Mississippi in 1844 to escape the bad health of the canebrake, and Mrs. Howe had to raise her children alone without her female relatives.[30]

Conflicts sometimes undermined these relationships, just as they marred relationships among male kinfolk. Mississippian Ann Archer disliked a cousin whose manners were not "cordial" and

John Tayloe III
Courtesy of the National Museum of Racing and Hall of Fame

Henry A. Tayloe
This photograph was taken when he was an elderly man.
Courtesy of Huntington Boyd

John J. Ambler, Sr.
Valentine Museum,
Richmond, Virginia

John J. Ambler, Jr.
Courtesy of the Virginia
Museum of Fine Arts

John C. Calhoun
Courtesy of the South Caroliniana Library, University of South Carolina

Andrew P. Calhoun
Courtesy of the South Caroliniana Library, University of South Carolina

Stephen D. Miller
Courtesy of the South Caroliniana Library, University of South Carolina

James Lide
Courtesy of the South Caroliniana Library, University of South Carolina

Eliza Townes Blassingame
Courtesy of India Earle Pepper

Rachel Stokes Townes
Courtesy of the Baptist Historical Collection, Furman University Library

Israel Pickens
Courtesy of the Alabama Department of Archives and History, Montgomery

Martha Lenoir Pickens
Courtesy of the Alabama Department of Archives and History, Montgomery

Leonidas Polk
Courtesy of the Jessie Ball DuPont Library, University of the South

Frances Devereaux Polk
Courtesy of the Jessie Ball DuPont Library, University of the South

Branch T. Archer
Courtesy of the General Libraries, University of Texas at Austin

believed "it is not in her nature" to treat relatives with the proper kindness, but she, like most women, tried to preserve good relations whenever possible. Mrs. Archer included her disagreeable cousin in family dinners, and another plantation mistress felt obligated to invite her "melancholy" cousins to visit. Adelaide Crain strove to build a good relationship with her new sister-in-law because of the "tie of kindred" that now existed between them.[31]

When planter women re-created interdependent relationships with kinfolk, it was an accidental result of the settlement process, and it happened whenever men chose to settle near kinfolk. Most women, however, could not reestablish the array of relationships they had enjoyed in the seaboard because they had so little control over their mobility and because men did not habitually assist them in their travels. Women, who wanted so badly to preserve familial relationships, could rarely do so to their satisfaction, while men, who initially wanted to abandon these ties, were able to resurrect them. This outcome had its roots in the different ways that the sexes utilized kinship and in the comparative powerlessness of women within the planter family.

Material Success and Social Class

Planters' sons had a mixed record in their quest for fortune in the new land, and in the acquisition of slaves they did no better than their brothers who remained at home. Table 2 lists the slaveholdings of fathers and sons in twenty families in which the father owned at least twenty slaves and had a minimum of two sons, at least one of whom migrated from the seaboard to the Southwest. These twenty fathers had a total of ninety-four sons, eighty-six of whom lived to adulthood; they fall into two groups, forty-nine who migrated to the frontier, and thirty-seven who stayed in the seaboard. Twenty-six of the forty-nine migrants (or 53 percent) owned at least twenty slaves at some point in their adult lives before 1860; twenty-four of the thirty-seven men who remained in the seaboard (65 percent) also joined the planter class at some point. In addition, four men who became planters moved to the Southwest and then returned east, so they fall into neither category. These findings, while based on a small number of families, suggest that men who wanted to become planters may have been wiser to remain at home, despite the apparent soil exhaustion. Their uninterrupted participation in the kin-based economic systems of the seaboard may have made the difference; some also had access to seaboard banks, which were more stable to begin with and rebounded from the crisis in 1837 more quickly than Southwestern banks.

Furthermore, twenty-four of these fifty-four sons who became planters also equaled or surpassed their fathers' slaveholdings at some point by 1860; among these men twelve were migrants, eleven remained in the seaboard, and one was a return migrant. These twelve highly successful migrants constituted 24 percent of migrants; the eleven highly successful seaboard residents constituted 30 percent of that group. Once again, planters did somewhat better over the long run if they remained in the seaboard.[32]

The twelve migrants who equaled or surpassed their fathers were typical in many respects, most being natives of North Carolina who settled in Alabama in the 1830s. All were planters, although several had other careers in medicine, the law, and the ministry. These twelve men hailed from seven different families, and some men began their venture with an advantage: Leonidas and Andrew Polk both inherited large sums; Gaius, Needham, and William Whitfield inherited seven slaves each from their father; James M. Calhoun inherited seven slaves after he moved from South Carolina to Alabama. Three other fathers of highly successful migrants—Isaiah DuBose, Isaac Otey, and Samuel Pickens—died after their sons migrated, but they left no wills or probate records. In the absence of these documents, it is impossible to tell if inherited property was the key to the success of these twelve men.[33]

But family correspondence reveals that all of these men engaged in economic exchanges much like those of the seaboard. We know that the Lea sons received considerable assistance from their father when they moved west, and they continued to receive infusions of cash for years afterwards. Samuel Pickens of the Alabama Black Belt also received help from his brothers, cousins, and in-laws in the Southwest and the seaboard. Men in the Polk family of Tennessee and North Carolina also enlisted each other in their various planting ventures, as did the men in the Whitfield family of Alabama, Mississippi, and North Carolina.[34]

Many Southwesterners who failed to prosper did not engage in these systems of assistance. Burton Carr left Virginia enraged with his older brothers and moved restlessly throughout the Southwest, corresponding only with his younger brother. By 1850 he was working as a whiskey distiller in Barren County, Kentucky, where he owned a thousand dollars' worth of real estate, but he did not own a single slave, having lost along the way the only bondsman he inherited from his father in 1812. Many of his personal possessions were destroyed in a house fire in 1850, and his hopes for a quick fortune were dashed that same year when a local railroad company decided not to place a route through his property. Other migrants may have asked for help when it was too late, and some, like Samuel Townes, may have lost their health; yet others may have had Burton Carr's spectacularly bad luck. Many factors contributed to their difficulties, but the assistance of relatives made a key difference, perhaps the key difference, in the long-term wealth acquisition of these men.[35]

All of these findings suggest that Southern white men who reached adulthood during and after the 1830s may not have been as successful as their fathers had been. The percentage of slaveowning families across the South decreased from 36 percent in 1830 to 25 percent in 1860. By the 1850s wealth in land and slaves was highly concentrated, and slave prices were rising beyond the reach of many aspiring planters; the concentration of wealth may have been greatest in the Black Belt. Opportunity, then, did exist in the Southwest, but it was not quite that imagined by the exuberant migrants of the 1830s. Success required hard work, good luck, and the sustenance of male kinfolk. It was difficult, if not impossible, to live out the ideal of economic independence.[36]

Planter women could point to few positive results from the family venture in the new land. Marianne Gaillard, whose husband Thomas was in financial straits by the mid-1840s, moaned that "God brought us here to bring us down in the world" but reminded herself that many people were "worse off" than her family; she hoped that God would be merciful and spare them from bankruptcy. Mrs. Gaillard was the exception to the rule, however, for even the wives of prosperous planters had little or nothing to say about economics, probably because many were as uninformed about financial matters as seaboard women were.[37]

But what was more significant, women did not define success in material terms; to them the good life meant preserving ties with a large family, which was impossible to do in the new country. Some made their views clear when they expressed feelings of isolation, homelessness, and displacement even after years of residence in the Southwest. After six years in Tennessee, Frances D. Polk confided to a relative that she was "sometimes low spirited at the thought of being so much alone, & I find the reality bad enough thus far." Septima Rutledge, the wife of a wealthy planter, felt that she was in "exile" from her native South Carolina after eighteen years in Tennessee. Martha Gamble of Alabama, who had written to her cousin faithfully for over thirty years, still wished in 1860 that "you were so nigh that I could step over and see you any time I wished, but," as she bitterly remarked, "that privilege is denied me."[38]

Planter women also passed judgment on larger questions of geographic mobility and the pursuit of riches, implicitly criticizing male greed. Sarah Gayle pitied the genteel wife of a migrant who saw fit to "bury her" in an obscure Alabama town because "*money* money must be had." Mary Drake proclaimed that only slaves and "hardy industrious" white men, particularly those who loved money above all else, were suited for life in the region. Another planter's wife declared that she would not have left North Carolina for "all the wealth in Mississippi" because the company of her seaboard relatives "contributed more to my happiness than riches ever will." North Carolinian Evalina Lenoir, whose family had settled in Boone County, Missouri, in 1835, agreed. When she thought of the beloved relatives and friends she had left behind, she decided it was "too great a sacrifice for the sake of a little more of this world's goods."[39]

Despite these strong words, planter women had no choice but to accept the situation. Time and again they expressed feelings of sorrow, resentment, and anger in letters to each other but brought themselves up short when it came to challenging planter men directly or drawing more general conclusions about the distribution of power within the family. Evalina Lenoir told her sister-in-law in North Carolina that she believed they would never meet again. "Oh Mira this sentence has produced a serious pause," she wrote, as she realized that "a separation from friends I dearly love has proved to be my fate." "But enough," she admonished herself, and closed the letter. One young woman described her great loneliness in Mississippi to her mother, confessing that she still had "the strongest attachment" for her native North Carolina, but decided to drop the subject because it was "perhaps wrong for me to express my feelings in this way." Mary Drake, who was miserable in the Southwest, often imagined returning to North Carolina but checked herself, realizing the "fallacy of such thoughts." Instead she would try to "find the bright side of the picture (if indeed there is one) or try not to think at all." After much struggle, this was the solution many women reached: they tried to accept the circumstances of their lives, and if they could not accept them, they tried to deny them.[40]

A few women, however, enjoyed life in the new land, just as a few had supported the original decision to leave the seaboard. One such woman was Joanna Hall Townes, wife of Samuel Townes, who moved from South Carolina to Alabama in 1834. She was probably fleeing the family much as her husband was; she rarely went home to visit her relatives, and she wrote to her in-laws infrequently, but her marriage, unlike many others, seems to have flourished on the frontier. There is no surviving correspondence between Mr. and Mrs. Townes, but their letters to other people reveal that they respected each other and enjoyed each other's company. In 1834 Samuel told his brother that his "excellent" spirits were a "consequence of my improved health I suppose but Joanna insists that she has had the greatest hand in it," and he admitted that his happy marriage contributed to his outlook. She in turn reported that her husband was "entering actively into all the business of life and his profession with spirit, energy, perseverance, and pleasure." She predicted, in fact, that he would be "the *Patrick Henry* of Alabama."

Joanna Townes was blessed with tremendous physical energy, and she worked in her household with a will. Her husband boasted that she was "industrious, neat, and *economical.*" Proud of what she called the "rising prosperous people" of Alabama, she believed that merit was rewarded in its open society. In fact, she enjoyed its "stir and novelty" so much that a return to Carolina and the "even tenor of our old ways" would be "insupportable." She fell silent in the mid-1840s when her husband lost his fortune and his health, but she did not welcome the idea of resuming close relations with the seaboard family. Her brother-in-law Henry Townes related that she did not want to accept "charity" from the Townes family because it would place her own family in "an entirely dependent situation."[41]

Finally, planter men and women tended to respond differently to the breakdown of older standards of social status. Individuals made and lost fortunes quickly in a volatile economy, and social class was harder to assess in the Southwest; the benchmarks of kinship, which had meant a great deal in the seaboard, did not matter very much in a world filled with strangers. A few women, like Joanna Townes, celebrated the change, but the typical planter woman still wanted kinfolk to provide the foundation for her social networks. When M. E. Perkins believed that her family might move from Alabama to Florida, she tried to persuade her cousin to go with them so she could "help out our own society" in "that wild country."[42]

In the absence of kinfolk, women reached out to make friends at church services and during social calls, and they sought women who practiced the reciprocity that characterized kin relationships. When Evalina Lenoir first arrived on the Missouri frontier, she received "quite acceptable" gifts of food from three neighbors and gave them presents in return. After several women visited a plantation mistress in Alabama when her son was born, she called on them first "of course" when she was well enough to go out, "as those ladies were the first to call on me." When Texan Mary Maverick wished to compliment her neighbors, she said that they were as kind to her as kinfolk would have been.[43]

Many of these new friends must have been affluent, and the term "ladies" probably signified that they were members of the planter class. But women were typically more concerned with whether others behaved with the proper reciprocity than with wealth per se. A Tennesseean, for example, disapproved of her neighbors because they were "vain and silly" and did not observe the principles of reciprocity, "never making themselves of use or assistance," and Adelaide Crain disliked the shallow, ostentatious women who lived near her Louisiana home. She was put off by their "gaudy" attire and "gay plumes," and when she saw them "flitting" past her window she missed her female relatives in North Carolina more than ever.[44]

Men took a somewhat different view of kinship and social status, disdaining individuals who thought a family name was enough. A Mississippian was scornful of the many people who claimed to be related to the famous Polk family. Even Samuel Townes, who had to turn back to his own family for help, ridiculed Virginians who claimed to be from the "first families," which was a matter of "questionable respectability" in his eyes. "Personal merit constitutes the only claim to character and consideration," he believed.[45]

Wealth and professional success, rather than kinship in and of itself, were more important to most men in assessing social status. Henry Tayloe was impressed when a lawyer and congressman moved to his Black Belt neighborhood, and another settler was delighted by the arrival of several very wealthy planters. Men, like women, expected their friends to observe the principles of reciprocity, but they wanted financial aid, not companionship or social calls. A Louisianan proclaimed that his new friends would "go purse in hand with me" and help him buy livestock.[46]

An incident between planter William Gale and his mother Ann Gale highlights these differences between the sexes. When Mrs. Gale, a middle-aged plantation mistress from Tennessee, visited her adult son William in Mississippi in 1844, she paid a social call on his overseer's wife, a woman who was not related to her. As a parting gift, Mrs. Gale presented the woman with a bolt of calico cloth. William Gale erupted when he discovered what she had done, storming that her gesture was a sign of "*intimacy* and *equality*" that endangered the family's "station." He knew that his mother would disagree and anticipated her reaction: "too much pride—too much pride my son!" He demanded nonetheless that she never visit the overseer's wife again.[47]

The planter family in the Southwest proved to be capable of reestablishing some of its functions, such as material assistance among its male members, but it succeeded only partially at best in re-creating effective kinship relations among female relatives; furthermore, kinship no longer served as a principal indicator of social status. Other traditional functions of the family, especially those concerned with child-rearing, failed because of the nuclear family's isolation from kinship networks, the different agendas of its male and female members, and the distinctive sex roles of the Southwest. Furthermore, the family's geographic isolation allowed planter men to engage in behavior that the presence of kinfolk had moderated and held in check, with unhappy consequences for planter women and slaves.

5

To Live Like Fighting Cocks: Independence, Sex Roles, and Slavery

New kinds of behavior appeared among the sons and daughters of planter migrants, resulting in part from the disintegration of kinship networks, and at the same time sex roles for adult men and women changed in unforeseen ways. Many planter men adopted more aggressive, self-absorbed forms of behavior, embracing the psychological aspects of "independence" whether or not they achieved economic autonomy, while many planter women grew even more dependent on men than they had been in the seaboard. Sex roles became extreme versions, almost caricatures, of the ideals of male independence and female dependence that seaboard men articulated in the 1830s. Not everyone engaged in these behaviors, but sex roles in the Southwest were clearly evolving in new directions.

Changes in sex roles affected race relations too, as many planter men broke the pact of paternalism and tended to equate masculinity with mistreating slaves. Many planter women, however, tended to abide by the precepts of paternalism and related to slaves much as they had in the seaboard. In fact, the changing conditions of planter women's lives may have allowed a few to sympathize with slaves as individual human beings.[1]

Child-rearing and Kinship

Many people frequently remarked on what one writer, an anonymous "Mother," called the "rudeness and ill-behavior of children in this latitude," but the sexes did not agree on the cause of the problem, or even on whether it was a problem. Many women, like the anonymous writer, worried about their children's behavior. Sarah Gayle believed that the training of other relatives was crucial in rearing young children, and she feared what would happen to her daughters if she died with "no mother, sister, aunt or niece of mine to watch over and guard them while they are rearing up to womanhood." These relatives would have formed their principles with "untiring zeal" and taught them "modesty," "goodness," and "purity." She also wished "it had been permitted for my children to have been nursed and loved by their grandmother" in her native South Carolina. Seaboard women were alert to the potential dangers of allowing children to grow up without their kinfolk. Caroline Gordon observed that her Southwestern cousins in the James Harvey Gordon family lacked the social graces and emotional stability of their mother, a native of the seaboard. Whenever Southwestern women could persuade men to arrange it, they sent their daughters to visit relatives in order to introduce them to the proper values.[2]

But it was most often young men who were censured as wild, disrespectful, and rebellious. Frederick Law Olmsted heard many stories about wild planters' sons, and he encountered a crowd of them "drinking, smoking, chewing, and betting" in a Natchez hotel. Another planter's son, who had the manner of a "pert, forward puppy," boasted to a visitor that he could sleep late every day because his family was rich. A Virginian who visited his younger brother in Mississippi reported that his sibling no longer listened to his advice and "thinks himself quite a man."[3]

Editorialists in Southwestern newspapers declared that parental authority had somehow diminished, but they disagreed as to who was at fault: some blamed parents, and others blamed sons. The most sophisticated discussion came from an anonymous article in *DeBow's Review* in which the writer, probably the magazine's editor James DeBow, contrasted the seaboard family with its Southwestern counterpart. In the seaboard "as long as the parent lived, the child was not free from his control." In the new country, however, "we have gone to the other extreme" in rejecting parental authority, "carrying to excess our notions of liberty and freedom from restraint." Southwestern parents were partly at fault for allowing "the child, at an early age, . . . [to throw] off all control."[4]

As the editorialist suggested, some fathers encouraged this kind of behavior. Many men who migrated in their twenties and thirties remained committed to the values of "manly independence" when they became fathers, and they rewarded signs of aggression in their sons. A Mississippian was delighted that his son threw tantrums "regardless of chairs, tables, or danger of breaking his nose," and he concluded with satisfaction that he was "a great boy." Many fathers who did not achieve true economic independence nonetheless taught their young sons the *psychology* of independence. Samuel Townes, whose finances declined so swiftly in the 1840s, described one of his sons in 1843 as "a perfect tinder box," a "devil" with an explosive temper, whom he approved of as "a noble boy." Andrew P. Calhoun, who was never able to pay his debts, called his son Duff "the most independent chap you ever saw," adding, "He will have energy enough to keep pace with the foremost in this stirring country of ours, where he 'who lags is lost.'"[5]

Women, however, worried a great deal about their sons' behavior. Joanna Townes, who was usually so loyal to her husband, Samuel, thought that her wild son was "full of [him]self" and lacked character. Eliza Townes Blassingame, Joanna Townes's sister-in-law, despaired over her teenaged son John, who drank to excess, refused to work, and got into fights with "brutes" in Perry County, Alabama. Sarah Gayle of Greensborough, Alabama, was anxious about her disobedient, "sullen" son Matthew, whose father, John Gayle, encouraged the traits of independence in all of their children.[6]

These problems many women blamed on the absence of kinfolk and, by implication, on the men who brought them away from their relatives. Regarding a "wild" young kinsman, Sarah Gayle commented that "subsequent care and culture" could not overcome "the pernicious effects" of "early indulgence." Seaboard women were just as critical of the upbringing of their frontier relatives. South Carolinian Anne Dent was disgusted by her spoiled, impulsive grandsons, the sons of John Horry Dent, who grew up in Alabama in the 1840s and 1850s. After several visits to the household, she told John Horry to send them away to school to "relieve the house from the torment of two great idle boys." She believed that all of his children had suffered because they had been reared with little contact with their relatives, and he should "now feel the value and regret the loss" of their influence. At least one man visiting from the seaboard concurred. A Virginian believed that the "rough, rude, untutored" sons of a Texas couple had been in various kinds of trouble because the family was so isolated from their relatives.[7]

Whenever possible, women tried to reform their sons by placing them with appropriate male kinfolk. Eliza Townes Blassingame turned to a series of relatives for assistance in dealing with her son John. His father, William, was little help; in fact, Joanna Townes believed that John's problems stemmed from his father's "mismanagement." His mother wanted to send him to live with an uncle in Alabama who was known for his "strictness and proper management," but that experiment failed. After William Blassingame's death in 1841, Mrs. Blassingame dispatched John to in-laws in South Carolina, but he continued to drink and misbehave, denouncing his seaboard relatives (with the exception of an uncle who lent him some money) and came back to Alabama, bringing "shame, trouble, and mortification" upon his mother and sisters. In this family, as in many others, sporadic contact with kinfolk was insufficient to reform unruly sons.[8]

Independence and the New Masculinity

Many adult men also began behaving in new ways as sex roles were transformed. They embraced the psychology of "manly independence" that had coalesced in the seaboard of the 1820s and 1830s, and they expressed it to the full in the Southwest. Such manly independence manifested itself in a kind of daredevil masculinity, with an emphasis on prodigious drinking, gambling, skill with a gun, and, perhaps, less clandestine sexual activity with slave women, while a man's obligations to his family became more tenuous. Samuel Townes, who articulated so many of his generation's goals, summed up the new masculinity when he appealed to a brother to move west in 1834 because "you can live like a fighting cock with us"—a striking image connoting strength and combativeness. He urged another brother to leave the seaboard in 1841 because he had too much talent "to hang around Mother and drivel away your life." This is what men wanted: to live a defiantly autonomous life free from the interference of the family.[9]

Men drank heavily in all parts of the South and throughout antebellum America, but the consumption of alcohol among Southwestern planters seems to have been enormous. Simon Phillips, who lived as a slave in Alabama's Black Belt, recalled that almost every plantation had a whiskey still on it. Young planter men seemed especially likely to take up these habits of consumption, drinking to get the approval of their friends and boasting about their "manly independence." According to one resident, the typical young man was soon depraved, his "desires and appetites . . . unrestrained," a source of misery to his family.[10]

Indeed, family correspondence throughout the period is filled with accounts of the heartache caused by heavy drinking and outright alcoholism. One planter tried to help his brother get started in some kind of business, but he wasted one opportunity after another because he was a "*drunkard.*" An Alabaman related that his alcoholic son, who sometimes worked as a schoolteacher, never wrote to the family; he heard occasional reports about him from one of his in-laws. Jane Polk was alarmed by the drinking bouts of her son, Franklin, one of which was so severe that it "like to a killed him." Franklin promised his mother that he would stop, but he did not, or could not, and alcoholism probably contributed to his death in 1831 at age twenty-eight. Alcoholism ravaged other planter families in the Southwest. Samuel Townes's brother-in-law, William Blassingame, drank himself to death in Alabama, as did Thomas J. Calhoun of Mississippi, the brother-in-law of Samuel's brother Henry, and two cousins of Alabaman Ann Finley. The relationship between drinking patterns and culture is necessarily complex, but many Southwesterners believed that something in their way of life promoted destructive drinking.[11]

Planter men also took up gambling with zest, playing at Natchez and New Orleans and at springs, resorts, and racetracks throughout the region. Some bred horses for racing, a sport that had long been associated with wealth, power, and leisure. Samuel

Townes and Henry Tayloe both brought thoroughbreds with them from the seaboard and were passionate followers of the turf. Planters gathered in gambling dens, which went by the simple but evocative term "hells," where disputes over card games sometimes ended in violence. In these places one planter watched the "degradation" of young men who drank and gambled until they were "drawn in the vortex and overwhelmed by disgrace."[12]

Some gambling dens, such as the Bell Tavern in Cahaba, Alabama, were reserved for elite men. It was the favorite haunt of many Black Belt lawyers and planters, who sometimes spent days playing billiards or poker, betting thousands of dollars on the turn of a card, and many slaves changed hands to pay their debts. Migration to the Southwest was itself a gamble, so it is perhaps not surprising that these men would be drawn to games of chance.[13]

Planter men also took part in ritualized violence, preeminently the duel. Affairs of honor took place all over the South, but duels reflected especially closely the aggressive masculinity of the sex role emerging in the Southwest. The congregation of single young men in the towns of the Mississippi Delta created an explosive social atmosphere that contributed to many duels. "High-spirited" young men gathered in public places, talking and lounging in hotels, cafes, and gambling saloons. Among them the code of honor was "woven of the finest texture and of the most sensitive materials," and it was "constantly appealed to" and rigidly observed.[14]

Alexander McClung, one of the South's most famous duelists, was a man of brooding spirit. A Virginian from a distinguished family, he settled in Jackson, Mississippi, by the 1830s and by the mid-1850s had fought in at least fourteen duels and killed ten men. McClung never married, and a friend described him as "wild" and "untameable"; he also gambled and drank. In 1855, when McClung was in his mid-forties, he shot himself to death in a Jackson hotel. Other duelists were not so self-destructive, but they showed the same obsession with honor, reputation, and proving their masculinity. William H. Polk, the moody younger brother of James K. Polk, moved to Mississippi from North Carolina in the mid-1830s and killed a man in a duel in 1839.[15]

Planter men also engaged in acts of random violence, the stabbings, shootings, and attempted murders that shocked travelers in the Black Belt in the 1830s. The threat of violence lurked beneath the surface of ordinary life as it did not in quite the same way in the seaboard, and bloody, unforgettable scenes confronted residents as they went about their daily business. John Blassingame, nephew of Samuel Townes, once got the worst of a fight when his opponent smashed a brick against his skull; he was carried through the streets of Marion, Alabama, on a stretcher, bleeding profusely, to his uncle's home. The cruelty and unpredictability of life in the Southwest is graphically represented in its regional humor, with its emphasis on dares, fist fights, and brutal practical jokes.[16]

More frequent sexual relations with slave women may have accompanied these other behaviors, according to scattered comments in planter correspondence. Young men occasionally confided in each other, and they boasted of these relationships. One planter's son bragged to his brother, "I have just returned from unchaste enjoyment of pleasant sensations in associating with my dear Jane." An Alabaman joked that he stopped his "love visits" to "the quarter" because he was seriously courting a white woman. Many years later, former slaves condemned the profligacy of their masters, especially unmarried masters. Adeline Marshall recalled that her owner fathered many mulatto children with a house slave on his Texas plantation, and another ex-slave, James Green, charged that masters and overseers forced sexual relations on slave women at will.[17]

Married planter men also had sexual relationships with slave women, of course, and they seem to have flaunted their involvements more than did men in the seaboard. When a Mississippi planter insisted on bringing his slave mistress into his home as a servant, his kinsman grimly remarked, "I certainly do not envy his family their happiness." An Alabaman also brought a mulatto woman into his home until his wife and scandalized neighbors forced him to sell her to New Orleans. Sarah Gayle was outraged by the "beastly passions" of white men who fathered children among their slaves and then sold their offspring like so many pieces of livestock.[18]

In this environment slave women may have been even more vulnerable to sexual abuse than they were in other regions of the South. Once again the evidence is sparse, but the story of physician William J. Wilson may be instructive. He became notorious in

Grimes County, Texas, in the late 1840s for harassing and raping white and black women. Among his victims was a slave woman named Bet; the doctor raped her when she was sleeping in the same room with one of his patients, and she became pregnant. Wilson left the area in 1850, but there is no record of her fate or the fate of her child.[19]

Another key component of masculinity changed in the Southwest: manhood was no longer automatically associated with duty to the white family, as it had been among older generations of seaboard men. Texan Solomon Page demonstrated that familial duties mattered less to him than the opinions of other men. He rode into a nearby town one day to buy food for his family but decided instead to join the army and fight in the Mexican War because he did not want other men to think he was a coward. Before he departed he told his wife, Harriet, "You will have to do the best you can." She replied that she hoped he would be killed for leaving his wife and children alone "in this wilderness." Southwestern men also tried to avoid associations with their female relatives in the seaboard. An Alabaman had not yet made his fortune but would not return to his native North Carolina because, as he told his cousin, "I know that I would not be satisfied to live with mother and sisters in that country."[20]

The career of Branch T. Archer of Virginia richly illustrates this disregard for familial duty, as well as other aspects of the new male role. The son of Peter F. Archer, a planter in Powhatan County, Virginia, he married Eloise Clarke in the 1820s and had at least six children. He left the state in 1831 after he killed one of his cousins in a duel and lived for many years in the town of Velasco, Texas, where he practiced medicine. He also helped found the Republic of Texas, and Archer's fiery personality, impulsiveness, and creative profanity made him a local legend. When an acquaintance from Virginia encountered him in New Orleans, he said that Archer "talks too much and too loud." After over twenty years in Texas, Archer still adhered to the Southwestern style of masculinity and lived it to the hilt. In 1854, at age sixty-four, he told a relative that "my nature is restless" and that action for its own sake, regardless of the outcome, was "necessary to my happiness." He once invited a relative to visit and proposed a steady diet of fish, fowl, and whiskey. "We can live like . . . fighting cocks," he proclaimed, echoing the image Samuel Townes used to describe this way of life.

Archer did not achieve economic independence, however, and he did not equal or surpass his father's slaveholdings of twenty-three slaves in 1820. He owned no slaves at all in 1850 and only forty-four hundred dollars' worth of real estate in Brazoria County, Texas. When Archer died in 1856 he had been living without benefit of matrimony for several years with Sarah Groce Wharton, the wealthy widow of William H. Wharton, another founder of the Texas Republic. His wife and five of his six children had either died, deserted him, or been abandoned by him.

His only surviving son, Powhatan Archer, followed in his footsteps, displaying many of the psychological attributes of a Southwestern man, if not the economic independence. He too lived in eastern Texas and practiced medicine. In 1855, when Powhatan was in his mid-thirties, he visited his cousins in Mississippi, and he did not make a good impression. His hostess, Ann Archer, confided to a relative that he "drinks very hard and was very disagreeable to us." She speculated that he was "addicted to drinking" and thought it was a pity that her cousin's son was so "worthless." In 1860 Powhatan lived in a boarding house in Brazoria County, a bachelor with property holdings of eleven hundred dollars.[21]

As these examples suggest, the behavioral attributes of "manly independence" developed apace with their own momentum despite the failure of many men to achieve economic autonomy. J. W. Calvert, a Virginian who migrated to the Southwest, told his cousin that he had made and lost a great deal of money over the years but declared, "I have always felt independent." In a similar vein, Isaac Cloud celebrated his "independent life" on the frontiers of Kentucky and Missouri, although he owned fewer than ten slaves.[22]

Furthermore, the economic leverage of more prosperous relatives was usually not sufficient to halt these transformations. South Carolinian Henry Townes, the conservative older brother of Samuel, was very much troubled by the drinking habits his nephew John Blassingame acquired during his teenaged years in Alabama, which had reached the point where John's "honor" was "at stake." Henry could not understand why his nephew, who was the oldest son, would not devote himself to caring for his family as Henry had devoted himself to his own family. Although Henry lent money to young Blassingame, his power as a creditor was not enough to affect the day-to-day behavior of his nephew hundreds of miles away.[23]

Independence and Femininity

Women reacted with fear, bewilderment, and anger to these changes in male sex roles. Some of the strongest condemnations came from Sarah Gayle, wife of a lawyer and politician in Greensborough, Alabama, as she recorded in her diary the consequences of the new masculinity in the marriages of her neighbors and relatives. Mr. Chapman drank so much that his wife returned with their children to live with her father; Mr. Clarke, recently admitted to the Methodist church, beat his wife with a piece of cowhide; Mr. Erwin neglected his wife, who was lonely for her relatives in Kentucky. In her own family, one of her in-laws, Maria Gayle, married a man who turned out to be a scoundrel; eventually Maria's cousin Benjamin Gayle murdered the man just as he was preparing to take the family to Mississippi. Long passages of the diary have been scratched out, probably by one of Sarah Gayle's descendants, but the surrounding text seems to indicate that these passages describe yet other conflicts within the Gayle family and between residents in the village of Greensborough. We can only wonder exactly what stories the stricken passages contain.

Sarah Gayle sympathized with these women, but she knew that they had no choice but to endure, especially when they had no kinfolk nearby to protect them. She believed that Mrs. Erwin, for instance, should try "scrupulously" to please her undeserving husband because she was "wholly dependent upon his care." Gayle realized that women were newly dependent upon men and more vulnerable without their kinfolk. One of her neighbors had given "her destiny up" to a husband who drank and gambled, and one of her aunts, whose spouse had become a "confirmed drunkard," lay dying alone from consumption because "one after another of her family . . . has dropped off."[24]

As Sarah Gayle intimated, female roles also changed in the Southwest, so that women became even more dependent on men than they had been in the seaboard. It is true that one aspect of their lives, their legal rights, did improve. Mississippi enacted the nation's first Married Woman's Property Act in 1839, and other Southwestern states passed similar statutes, while the seaboard states did not recognize these rights until the Reconstruction period. But legislators wrote these bills to protect men from creditors in the wake of the Panic of 1837 rather than to advance women's rights; the Mississippi law stipulated that the husband retained profits generated by any slaves owned by his wife. To make matters worse, judges interpreted the law narrowly when cases involving married women's rights came to trial. Furthermore, the social and material conditions of life in the Southwest, which affected women's daily lives more than these legal changes, were more repressive than in the seaboard. Women's work was more difficult; the environment they lived in was more dangerous; and, most important of all, they were isolated from their female kin who had been sources of companionship, strength, and assistance.[25]

Furthermore, transformations in sex roles increased conflicts and tensions in married life. Sexual relationships between planter men and slave women estranged planter husbands and wives, and they also troubled relationships between planter men and their mothers. Anne Dent told her son John Horry that if he allowed her grandsons to grow up in close proximity to slaves that it would end in "pollution of all kinds." The intemperance so prevalent on the frontier also created marital problems. In the village of Greensborough, Sarah Gayle observed several marriages deteriorating because of what she called the "beastly habit of drinking" among the husbands. Young children sometimes sensed that something was wrong between their parents. Kimbrough DuBose, who migrated from South Carolina to Marengo County, Alabama, drank every day from a bottle in his dining room. One day, as he prepared to take his toddy, his three-year-old son, Eugene, rushed up behind him, "pounded him furiously in the back, crying in his anger: 'Come out my mama's sideboard sir! Come out, I say!'" Eugene's older brother John, who witnessed the incident, recalled that "his rage was intense and his fight earnest."[26]

Some husbands enhanced their authority by inspiring fear within the household, such as John B. Dabney's kinsman who migrated from Virginia to the frontier. Dabney related that the man's anger was ferocious, like that of a "lion in his lair," whenever he perceived an insult. Although he loved his wife, he never allowed any challenge to his authority in the household, which would "be sure to bring down on the offender his hottest displeasure." His will was absolute, and he expected complete obedience from all. Dabney added lamely that he believed his uncle did not "abuse his power," but the man's wife and children have left no account of the family.[27]

We can see these trends at their worst in the marriage of William Blassingame and Eliza Townes Blassingame, the brother-in-law and sister of Samuel Townes. The couple married in 1818 in Greenville, South Carolina, when he was twenty and she was sixteen. Blassingame was the scion of a prominent family, and in the 1820s he became the town sheriff; by 1830 he owned over a thousand acres of land and about thirty slaves. None of the Blassingames' correspondence has survived, but Samuel Townes's letters, an admittedly biased source, show his sister to be a voluble, expressive woman; his brother-in-law comes off as icy, arrogant, and unpredictable. The couple bickered a good deal, and Blassingame was on bad terms with his in-laws, partly because the Townes brothers suspected that he lied about his property holdings during the settlement of their father's will in the late 1820s. But the presence of these kinfolk nonetheless gave the marriage some stability. The Townes brothers kept a close watch on their brother-in-law, and Eliza frequently visited her mother, to whom she was devoted. The equilibrium in the marriage shattered, however, after the Blassingames moved to Perry County, Alabama, in 1833. Samuel Townes settled the next year in Marion, sixteen miles from their plantation, and over the next seven years he recorded his sister's increasingly desperate plight.

William Blassingame drank heavily and began to disappear from home for long periods of time. His wife had to beg him for money to feed and clothe their six children, although he was a rich man, the owner of many slaves. When he was at home he began verbally and physically abusing his wife, and Samuel suspected he was having sexual relations with slave women. Townes began to despise his brother-in-law, whom he called a "dog," a "beast," and a "villain." After an incident in 1834 in which Blassingame struck his wife, Samuel and one of his brothers (who was visiting from

South Carolina) gave him a flogging and then boxed his ears. For several months afterwards Blassingame behaved himself, "fawning" on his brother-in-law, but then returned to his abusive ways. Meanwhile, he began having financial problems. He borrowed money from a state bank but by 1839 had sunk so far into debt that he feared he would have to flee the state to escape his creditors.

Samuel Townes loved and sincerely pitied his sister, who had never wanted to move to Alabama. Eliza Blassingame visited her brother whenever she could and was grateful for his protection, but she missed her mother acutely; she "cried like a whipt [*sic*] child" when Mrs. Townes canceled a visit to Alabama in 1841. Samuel was sometimes too preoccupied with his own worries to look out for her, and the sixteen miles between their households made visits inconvenient. He also knew that despite his best efforts to be "father, mother, brother, and sister" to her, he alone was unable to check Blassingame's behavior. Samuel came to believe that Eliza never should have been taken so far from their mother, who could have given her emotional support and guidance. Townes concluded that only death could end Eliza's suffering and was openly relieved when his brother-in-law died, most likely of cirrhosis of the liver, in 1841. The Blassingames probably would have been unhappy if they had remained in the seaboard, but their isolation on the frontier worsened William's destructive behavior and his wife's chronic unhappiness. If Samuel Townes realized that this was the ultimate outcome of living like a fighting cock, he did not discuss it in his letters.[28]

Other planter women who found themselves in troubled marriages turned to their kinship networks, when they were available, for protection and assistance. Like Eliza Blassingame, they took refuge with their relatives, making lengthy visits or inviting kin to live with them to help defuse tension. When marriages failed altogether, women also turned to their relatives. A young Virginian who migrated with her husband to Florida found that her spouse neglected and then abandoned her on the frontier; she finally wrote to her brother, who agreed to help her. But because women were so immobile, even contented wives realized that their happiness was bound up exclusively with their husbands if no relatives lived nearby. One of Sarah Gayle's friends, who was alone with her husband on the frontier, pointed out that their situations were "peculiarly similar, neither having parents, brother or sister, and both depending upon our husbands for our share of earthly happiness."[29]

The trials of Harriet M. Page show what could happen to a woman who became completely isolated from her kinfolk. When she and husband Solomon Page moved to Texas in the 1840s, they settled near her father and brother in Brazoria. Her husband was a compulsive gambler and forfeited many of their belongings in all-night card games, but her male relatives kept the household running. Her brother usually replaced the items her husband lost, and her father offered to give her property so she could support herself, but her husband would not allow her to accept it. When Mr. Page moved the family twenty miles away to a shack on the prairies, she was utterly alone. He once left her and the children on the prairie for over a week, and Mrs. Page feared that they would starve. "Oh the terrible inaction," she recalled, "when my little ones fretted with hunger, and day after day the sun rose and shone on the prairie empty of human life." They were saved when a woman Harriet Page had befriended decided to send a local minister to check on their safety. When the family boarded his wagon to return to "civilization," Mrs. Page looked back at the little house "till it became a mere speck and was swallowed up in the great wide prairie. So, I hoped, would its terrible memory fade from my life."[30]

Independence and Slavery

The ongoing changes in masculinity, especially the emphasis on individual autonomy and a man's freedom from obligations to dependents, contributed to significant changes in ideology and behavior regarding slavery. They probably influenced the writing of Josiah Nott of Alabama, who championed so-called scientific racism and proposed a vehemently racist defense of slavery in the 1840s and 1850s. Seaboard writers had defended slavery by comparing it to the family, but Nott broke with the assumptions underlying paternalism. Instead he formulated new theories about the origins of humankind and argued that black people belonged to an entirely different species and were less than human.

Nott's views gained him an international reputation, and other scholars describe how he became recognized as perhaps the South's chief proponent of slavery, but the link between his ideas and the changes in male sex roles has yet to be explored. Among the major proslavery theorists, he alone migrated to the Southwest and made a permanent home there, and he alone made no use of the family analogy. Unfortunately, few of his private papers, as opposed to his prolific public writings, have survived, but his biographical profile resembles that of many other migrants. A native of the South Carolina Piedmont, his father, Abraham, was a Yankee from Connecticut but his mother's relatives, the Mitchells and the Hendersons, had been in the state since the mid-eighteenth century, so that Josiah grew up surrounded by kinfolk in the planter gentry. He was a high-spirited young man, and all of his life people commented on his pugnacious manner. His father died in 1830, and after Josiah studied medicine in Columbia, South Carolina, and Paris, he settled in Mobile in 1836.

He did not cut ties with his family completely, as his biographer Reginald Horsman shows: he bought land with some of his relatives and took in a sister-in-law when she was widowed. But his letters to his male friends show that he was glad to be free of the dense social ties at home, where he had to visit a "large circle." He was exhilarated by the diversity of frontier life, and, despite his disclaimers, he clearly enjoyed the controversy his writings provoked, especially among clergymen who objected to his implicit criticisms of the Bible. Nott supplanted the paternalistic defense of slavery in part because he, like many white men in his generation, had rejected other pacts between generations of men and between the sexes that had governed family life in the seaboard.[31]

New kinds of behavior accompanied these shifts in the proslavery argument, as slaveowners in the Southwest became notorious for their harshness and their greed. Planter Richard Archer of Mississippi explicitly rejected the family analogy when he criticized a neighbor who treated his slaves as if they were "part of his white family." The spectrum of acceptable behavior expanded to include abusive behavior that was probably less widespread—although by no means absent—in the seaboard. A traveler believed that slaves in Mississippi were abused more badly than seaboard slaves, subject to the whims and passions of their owners. Thomas Brown, a Florida planter and native of Virginia, refused to hire slaves to his neighbors because they mistreated their own slaves and did not feed them properly.[32]

As was the case with other social changes, seaboard residents took note. Virginian William Tayloe scolded his younger brother Henry Tayloe for driving slaves so hard on his Alabama plantations. "There shall be no more cruelty," William Tayloe insisted, but his decrees had little impact on his brother hundreds of miles away. Henry Townes believed that slaveowners in South Carolina were "generally more civilized & humanized" than masters in Alabama, where he thought slavery, especially the hiring out of slaves, was a "cruel business."[33]

In his letters Henry Townes did not call his brother Samuel a cruel master, but Henry may have had him in mind. Samuel Townes drove his slaves very hard in the mid-1830s as he tried to reap profits from his Black Belt plantation. During the harvest of September 1834, Samuel's first year in Alabama, he thought that all the planters around him were on the verge of making fortunes, and the idea that he might be left out of the bonanza tormented him. He grew increasingly impatient with his overseer Mr. Tucker, whom he thought was not pushing the slaves hard enough, and he became so exasperated that he refrained from striking Tucker only "with difficulty." He was also very angry with his slaves Phillis and Marcellena, who picked 40-odd pounds of raw cotton a day while his other women slaves picked as many as 70. (Some of his male slaves picked as many as 190 pounds.)

Samuel Townes had inherited Phillis from his father in 1826 when she was still a girl, and she worked as his maid in Abbeville while he was studying law with Armistead Burt. She came to Alabama with him in 1834, and Rachel Townes, Samuel's mother, asked about both Phillis and Marcellena by name in a letter written from South Carolina that year. (Marcellena was probably originally bequeathed by Samuel Townes, Sr., to his son John in 1826.) These personal associations meant nothing to Samuel Townes, however. In a fury, he told his overseer to "make those bitches go to at least 100 [pounds] or whip them like the devil," and he threatened to go to the plantation and discipline them himself if Tucker would not do it. Marcellena's productivity almost doubled within the next few weeks, but she did not reach one hundred pounds. Townes apparently did not retract his blood-curdling order, and the overseer must have carried it out. Townes said nothing at all about Phillis's productivity, but the next year he sold her, complaining that she was not strong enough to work, she was nearly deaf, and she had lost her reason. The abuse that this vulnerable young woman suffered illustrates as few other incidents can the extreme cruelty of slavery on the frontier.[34]

When former slaves recalled their lives on the Southwestern frontier, they portrayed a regime that was, in the words of Texan Martin Jackson, full of "plenty of cruel suffering." William Oliver stated that "the cruelest treatment I know of in the United States . . . was done in the Southwestern states." Adeline Marshall declared that her master treated his animals better than he treated his slaves, forcing small children into field labor and beating adults to make them work harder. Andrew Goodman, another former slave, seemed to recognize that these transformations in slavery may have been related to changes in the white family. He recalled that a neighboring master "was mean to his own blood," beating his white, legitimate offspring as well as his slaves.[35]

Many planter men taught their sons by example and by precept to mistreat slaves. Samuel Townes, who must have struck terror in the hearts of his slaves, once proudly called his young son "the terror of the back yard little negroes." John Horry Dent allowed his sons to abuse slaves at will, and, according to Dent's mother, the boys were "at one moment associating as companions with a set of poor degraded slaves, and the next tyrannizing over them." A Louisiana planter praised his son, who was about ten years old, for whipping and cursing his slaves, and a bondsman remembered that the father laughed and commended the child as "a thorough-going boy." Boys began imitating adults at an early age. When Frederick Law Olmsted was traveling through Texas he observed an eight-year-old boy beating a puppy, crying, "I'll teach you who's the master."[36]

Many planter women, however, continued to practice the female version of the paternalism they had learned in the seaboard, and they often perceived slaves as human beings, as many planter men did not. Women described individual slaves working in the household and running errands more often than did men, and they mentioned slaves by name in their writings more frequently than men did, who usually referred to slaves as simply "the negroes" or "the hands." More frequently than men, they conveyed messages between slaves in letters to their own white relatives, usually their female relatives.[37] One planter's wife in Mississippi deliberately left space on the page to fill out for the slaves; she told her mother that the slave Erwin sent greetings to "'Pap,' Aunt Hannah, Mary, Emily, and Milly." Women seemed to understand the urgency of these messages as men did not. Marianne Gaillard of Claiborne County, Mississippi, told her brother in South Carolina that "Sukey begs you when you write again to let her hear from her Father & Mother—Snow & Binah—& her other relations."[38]

Planter women also appreciated hearing news about seaboard slaves, as men evidently did not. Elizabeth Blaetterman, for example, thanked her kinswoman for a letter which was "doubly interesting" because it contained information about her slaves in Virginia, but she was disappointed that she could find out nothing about several others she had freed before leaving the state. Seaboard women recognized that other women knew more about individual slaves than men did. Rachel Townes asked her daughter Eliza Blassingame rather than her son Samuel for information about their various slaves in Alabama, some of whom were owned by Samuel, not the Blassingames.[39]

The daily lives of planter women, who were immobilized on Southwestern plantations, brought them into sustained contact with slaves as those of planter men did not. Every day Adelaide Crain helped her house slaves serve breakfast (but not dinner or supper) and then met with one slave, Cora, to "scold," "expostulate," and "consult" about the family's meals. Ann E. Harris knew intimately the daily routines of her house slaves, in particular Harriet, who "goes to the spring the first thing in the morning, makes up the bed, clears the room, and sets the table." Her husband, James Harris, who wrote a paragraph at the bottom of the letter, did not mention slaves at all.[40]

Some planter women mistreated slaves, to be sure. Ann E. Harris once reported to her sister that she had whipped the slave Selina, who, she implied, had been more obedient back home, and other plantation mistresses no doubt punished slaves harshly. But abusing slaves was never part and parcel of the female sex role as it was of the male sex role, because women were not encouraged to demonstrate their prowess by dominating other people. Many slaves perceived mistresses as more approachable than masters, and some turned to mistresses to protect them from abuse. Former slave Lucretia Alexander recalled that she ran to the mistress on an Alabama plantation when the overseer tried to whip her, and "he knew he'd better not do nothin' then." James Lucas, another ex-slave who belonged to several masters in the Southwest, firmly believed that "[planter] wives made a big difference" because they "went about amongst de slaves a-lookin' after 'em," dispensing food, clothing, and medical care. He too thought that mistresses were more approachable, recalling, "When things went wrong de womens was all de time puttin' me up to tellin' de Mistis."[41]

In the correspondence examined for this study, no planter man wrote sympathetically about the work slaves performed, or described identifying with an individual slave, if only for a moment, as a planter woman did in a Louisiana household one winter evening in 1852. Adelaide Stokes Crain seems to have appreciated for the first time how hard slave women worked after she began running a household in Shreveport. Born in Wilkes County, North Carolina, in 1827, she was the youngest child of former governor and U.S. Senator Montfort Stokes, who deserted his wife, Rachel, and eleven children in 1831 to work as a federal Indian agent in Arkansas. Adelaide never mentioned his abandonment in her letters, but it may have increased her capacity to sympathize with human beings who felt overwhelmed by the circumstances of their lives. She was a sensitive, "brilliant" girl, according to a relative, and she was very close to her female cousins, especially Caroline Gordon. It was Adelaide Stokes who was delighted to spend the New Year's holiday in 1846 in a "time-worn mansion" surrounded by relatives.

In 1848 she married an ambitious lawyer, Lawrence Crain, and reluctantly accompanied him to the Louisiana frontier. He realized that a gracious, highly visible wife was an asset to his career, and he made many demands on her. Mrs. Crain had a full slate of social calls to make every week, in addition to caring for their small children, teaching them to read and write, doing household chores, and supervising house slaves. In the early 1850s, she was often homesick and sometimes felt that she could hardly cope with her new responsibilities. On November 11, 1852, she wrote Caroline Gordon that she had so much work to do that it made her head swim to think about it.

In the same letter Adelaide Crain recounted a recent conversation with Sophy, a slave cook, whom she found weeping in the kitchen after a large dinner party. When Crain asked her what was wrong, she cried, "Oh Miss Adelaide [,] . . . I have got so much work to do I can't do nothing." "Poor Negro," Crain wrote, "I can sympathize with her this week, for I have got so much to do that I spend half my time thinking what I shall do first." There were obvious limits to Crain's feeling and imagination—she sympathized with Sophy "this week" only—but she did inquire about Sophy's feelings, and she described the conversation to one of her closest female relatives. Even this brief moment of identification suggests that planter women continued to relate to slaves differently from planter men. When men compared themselves to slaves, they did it to dramatize their fear of dependency or what they perceived to be their exploitation, and they did not preface the analogy with remarks like "Poor Negro." Adelaide Crain acknowledged that slave women, like planter women, worked very hard and that they too sometimes felt exhausted by their duties; furthermore, she identified with an individual slave as a human being. But like Mrs. Brown of North Carolina, who almost compared herself to a slave after going to Tennessee against her will, she did not develop these fleeting insights to criticize slavery, the planter family, or the society in which she lived.[42]

Crain's biography suggests some important connections between sex roles and race relations, between changes in the family and changes in slavery. As the pact between planter men and planter women broke down, the pact between planter men and slaves disintegrated as well. While many planter women still tried to observe the tenets of paternalism, many planter men did not. In fact, many of these planter men abandoned the very concept of human relationships based upon obligation and reciprocity, which had done something to ameliorate the inequitable social relations in the seaboard. In the Southwest the worst aspects of a patriarchal, slave society came to the fore.

Conclusion

Thomas Sutpen became the richest man in Yoknapatawpha County, Mississippi, but his "design" failed, and his family failed. William Faulkner returns again and again to two themes, Sutpen's unbridled materialism and his insensitivity toward the women in his family. Sutpen believed that he could buy anything or anyone he wanted, and he measured human relationships in terms of "receipts." Yet his family destroyed itself in a nightmare of miscegenation, murder, and incestuous feeling brought on directly or indirectly by Thomas Sutpen's ruthless disregard for human relationships.

The lives of planters' sons who left the seaboard for the Southwest usually did not end in so much horror, but their search for independence was filled with unexpected outcomes, and many of them failed in their design. Those who came to the Southwest intending to establish their independence and reject the all-encompassing seaboard family found that they needed the family's resources in a volatile economy. Over the course of their adult lives, about one-half of them became planters, and one-quarter equaled or surpassed their fathers' slaveholdings. They expressed their independence nonetheless in a new male sex role, drinking, gambling, and fighting with a new sense of license, and they may also have engaged in sexual relations with slave women with a new sense of license. In their social relations with other whites, men reached for a more modern concept of society, in which status was determined by achieved traits such as wealth rather than by being a member of a particular family.

Planter women's experiences could not have been more different. They came to the Southwest looking backward; even though many men no longer abided by the pacts that helped the seaboard family function, many women clung to the seaboard model because they were trying to hold on to the only way of life they had known, as human beings often do when they are confronted with profound change. When women tried to forge friendships with other women, they used kinship relations as a model and were less concerned with an individual's wealth. If their male relatives settled near relatives, they were able to approximate viable kinship networks, but most women found themselves alone with their nuclear relatives on the frontier, where even the physical environment seemed strange and unlovely. They experienced all of the isolation of modernity with none of its easy geographic mobility. Freedom of movement was a privilege in the antebellum South, not a right, and most women's kinship networks deteriorated, just as they had feared. The final result of all these changes was that women became even more dependent on men than they had been at home.

These transformations in the structure of the family and in sex roles affected race relations in significant ways. Slaveowners in the seaboard first suggested the paternalistic model of slavery, and many of them practiced it there. But many of the young men who came to the Southwest rejected the pact implicit in the paternalistic model, just as they rejected the seaboard family, and turned toward a more deeply racist concept of slavery. They refused to recognize the humanity of slaves and assumed that they owed their slaves little or nothing, disregarding family ties within the slave population and forcing slaves to work in brutal conditions.

Many planter women, however, continued to practice their own version of paternalism. They still perceived slaves as human beings with close ties to cherished relatives of their own, and they acknowledged the destructive impact migration had on the slave family as well as the terrible living conditions slaves endured on the frontier. A few may have begun to identify with slaves, recognizing some disturbing parallels in their own experiences as involuntary migrants. These women did not criticize the institution of slavery as a whole, but neither were they its unfeeling supporters. Their paternalism, for all of its limitations, was preferable to the cruel treatment of slaves so prevalent on the Southwestern frontier.

This story of planter migration, then, highlights the divisions within Southern society: between generations of planter men, between planter men and planter women, and between family life and race relations in the seaboard and the Old Southwest. It also illuminates the high social and personal costs of migration, the deprivations of the planter women who took part in it, and the burdens of the slaves who helped settle the Southwestern frontier.

A Note on the Tables

Table 1: Household Structure among Planter Families, Selected Seaboard Counties, 1810, 1820, 1830

My findings are based on a random 10 percent sample of all adult white male household heads from these seaboard counties; I then used whatever planter men surfaced in that 10 percent sample. I selected rural, agricultural counties of the Piedmont with boundaries that remained fixed until 1860 and that contained the homes of at least one planter family whose papers I read. I excluded all female-headed households and all individuals living alone.

The federal censuses for the years 1810, 1820, and 1830 list the name of the head of each household and the gender and age categories of other household members, sufficient information to allow scholars to estimate familial relationships among household members. If a household included one adult male and one adult female who were of marriageable ages, or a couple with at least one child, I classified it as nuclear. If a household included "extra" adults beyond an adult male and female who appear to be parents, I classified it as complex. If a household included individuals whose age categories made it impossible to decide whether they belonged to the generation of parents or children, or some other unusual combination of individuals, I classified it as ambiguous.

Table 2: Slaveholdings of Migrants and Their Fathers and Brothers

I examined men's slaveholdings rather than women's because men owned most of the property in this society and because men left more records of their wealth. I selected slaveholding as an index of economic success because most men aspired to be planters, it can be measured (unlike holding political office), and it is recorded in every federal census (unlike landholding). The number of slaves a man owned tells only part of his financial history, of course; the gender, age, skill, and health of slaves also affected their value. The censuses give the gender and age categories of slaves through 1840, and in 1850 and 1860 they provide the gender and exact ages but still do not provide the names; it was impossible to identify and estimate the value of over two thousand slaves. The fact that slaves are not named also prevents scholars from discerning whether a man purchased new slaves or kept the same slaves between one census and the next. Nor did the census indicate whether a slave was owned outright or was on loan or hire from another slave-owner. So the number of slaves a man owned stands as only one measure—and an approximate measure—of his wealth.

The biographical profiles describe men who appear in the manuscript collections I read; they are a "subset" of the collections and are not statistical samples. I eliminated twenty-two families because I could not obtain sufficient information about them. For example, I dropped the family of South Carolinian Levi D. Wigfall, father of Louis T. Wigfall, because another son, Arthur, does not appear in a single census from 1820 through 1860, even though he was a prominent minister and resided in Edgefield County, South Carolina, the entire time. The experiences of these other migrants are nonetheless described in the text. There is a certain bias in the table toward rich families (such as the Whitfields) and the famous (such as the Polks), who generated more documents.

I reconstituted the profiles from a variety of sources, including manuscript correspondence, published correspondence, memoirs, diaries, genealogies, wills, obituaries, biographies, biographical dictionaries, county histories, state and federal census returns, and indexes to the census returns. The census returns, the main source for the table, have many imperfections: the censuses between 1790 and 1840 do not list all household members and give only one figure for the number of slaves owned by the head of the household; this makes it difficult to discover the slaveholdings of men other than the household head unless those slaveholdings are described in another source, such as family correspondence. (The censuses of 1850 and 1860 list slaves on a schedule separate from white households, naming individual slaveowners.) Furthermore, some heads of household are not listed, perhaps because they were not home when the census taker arrived; this may explain some of the mysterious gaps in the census, such as the absence of John Finley, Sr., from Wilkes County, North Carolina, in 1820. Entire censuses are also missing, such as the Tennessee census of 1810, which results in a serious underreporting of the slaveholdings of William Polk, Sr. (The Virginia censuses of 1790 and 1800 were destroyed, and the census of 1790 was reconstructed from tax lists that did not list slaves.)

Finally, the lack of detailed information in the censuses before 1850 makes it impossible to identify some individuals with certainty. For instance, four men named William Irby are listed in the Virginia census index of 1830, three of whom were the proper ages, but none of them lived in Nottoway County, where the "correct" William Irby actually lived. Other sources did not offer additional clues that would have allowed me to choose the correct man. In such cases I entered "no listing" in the table. Similarly, the Edmund Irby who is listed in the Virginia census of 1840 is between seventy and eighty years old, much too old to be the son of William Irby; therefore I entered "no listing" in the table.

As I compiled the slaveholdings for sons, I noted slaveholdings for men over the age of twenty. When the manuscript sources or the census indexes reveal that a man owned slaves in more than one county, I listed the slaves in all of the counties. The table contains 319 listings of slaveholdings, not counting the "no listing" entries.

Table 3: Sex Ratios among Slaves, Selected Seaboard Counties, 1820, 1830, 1840

The census of 1810 provides the total number of slaves for each slaveholder in each county but does not indicate the gender of slaves. Beginning in 1820, the census provides gender and age categories for slaves. I used the same age categories for both sexes.

Table 4: Sex Ratios among Slaves, Selected Southwestern Counties, 1830, 1840, 1850, 1860

I selected rural, agricultural counties that were popular destinations of migrants after 1830 and whose boundaries remained fixed until 1860. These counties also contained at least one family whose papers I read. The age categories in the census varied slightly: in 1830 and 1840 they ranged from age ten to fifty-five, but in 1850 and 1860 they ranged from age ten to fifty.

Table 5: Household Structure among Planter Families, Selected Southwestern Counties, 1840, 1850, 1860

Once again I took a random 10 percent sample of all adult white male household heads, and I used all planter men who appeared in that 10 percent sample, excluding all female-headed households and all individuals living alone. Because Texas joined the Union in 1845, only two of its censuses (1850 and 1860) are included. The census of 1840 provides the same level of information found in the censuses of 1820 and 1830, so I continued to use the same methods for classifying households.

The censuses of 1850 and 1860, however, provide more information: the gender, full names, exact ages, and occupations of each household member. I continued to use the methods of classification employed for Table 1, with these additional categories: if a household included two nuclear families living together, or if it included children with surnames different from that of the head of the household, or if it included adults other than the married couple who *shared* the surname of the couple, even if they were listed as boarders, tutors, or overseers ("extra" adults), I classified it as complex. If a household included a nuclear family and adults with surnames *different from* the head of the household who were listed as boarders, tutors, or overseers, I classified it as ambiguous if there was no way to determine whether these individuals were related to the "core" nuclear family.

Once again I included multiple slaveholdings whenever they could be ascertained.

Table 1. Household Structure among Planter Families, Selected Seaboard Counties, 1810, 1820, 1830

County	N	Nuclear	Complex	Ambiguous
Charlotte Co., Va.	38	39%	13%	47%
Powhatan Co., Va.	32	12	75	12
Bertie Co., N.C.	63	30	44	25
Nash Co., N.C.	14	14	57	28
Darlington Co., S.C.	26	23	27	50
Sumter Co., S.C.	59	29	25	46
Total N	232			
Percentage of total		27	38	35

Source: See "A Note on the Tables." N = number of households. Percentages are rounded in all tables, so totals may not equal 100.

Table 2. Slaveholdings of Migrants and Their Fathers and Brothers

1. JOHN J. AMBLER (1762–1836), planter
1790: census does not list slaveholdings.
1800: census destroyed.
1810: 13 sls., Frederick Co., Va.; 53 sls., Henrico Co.; 10 sls., city of Richmond, Va.
1820: 74 sls., Louisa Co., Va.; 34 sls., Hanover Co., Va.: 11 sls., city of Richmond, Va.
1830: 14 sls., city of Richmond, Va.
Three daughters, six sons:

Edward Ambler (1783–1846), planter
1810: no listing, Va. census.
1820: 52 sls., Henrico Co., Va.
1830: 18 sls., Culpepper Co., Va.
1840: 52 sls., Rappahannock Co., Va.

Thomas M. Ambler (1793–1875), planter
1820: no listing, Va. census.
1830: 47 sls., Fauquier Co., Va.
1840: 45 sls., Fauquier Co., Va.
1850: 40 sls., Fauquier Co., Va.
1860: 38 sls., Fauquier Co., Va.

*John J. Ambler, Jr. (1801–1854), planter and lawyer
Migrated from Virginia to Alabama in 1835; returned to Virginia c. 1837.

* = migrants treated in this study
sls. = slaves

1830: 57 sls., Amherst Co., Va.
1840: 17 sls., Madison Co., Va.
1850: 0 sls., Hampshire Co., Va.

Philip St. George Ambler (c. 1806–?), planter
1830: 47 sls., Amherst Co., Va.
1840: 64 sls., Amherst Co., Va.
1850: 48 sls., Amherst Co., Va.
1860: 89 sls., Amherst Co., Va.

*Richard C. Ambler (c. 1810–?), planter and physician
Migrated from Virginia to Alabama in 1835; returned to Virginia c. 1837.
1830: no listing, Va. census.
1840: no listing, Va. census.
1850: 39 sls., Fauquier Co., Va.
1860: 44 sls., Fauquier Co., Va.

William M. Ambler (1813–1890), planter and lawyer
1840: 42 sls., Louisa Co., Va.
1850: 51 sls., Louisa Co., Va.
1860: 82 sls., Louisa Co., Va.

2. PETER FIELD ARCHER (1756–1814), planter
1790: census does not list slaveholdings.
1800: census destroyed.
1810: 23 sls., Powhatan Co., Va.
1820: 28 sls., his estate, Powhatan Co., Va.
Four daughters, three sons:

William Archer, (before 1790–?) planter†
1810: no listing, Powhatan Co., Va.
1820: 34 sls., Powhatan Co., Va.
1830: 5 sls., Powhatan Co., Va.
1840: no listing, Powhatan Co., Va.
1850: no listing, Powhatan Co., Va.
1860: no listing, Powhatan Co., Va.

Peter Field Archer (before 1790–?), planter
1810: no listing, Powhatan Co., Va.
1820: no listing, Powhatan Co., Va.
1830: 19 sls., Powhatan Co., Va.
1840: 25 sls., Powhatan Co., Va.
1850: no listing, Powhatan Co., Va.
1860: no listing, Powhatan Co., Va.

*Branch T. Archer (1790–1856), physician
Migrated from Virginia to Texas in 1831.
1810: no listing, Va. census.

*Migrants
†Sons who equaled or surpassed their fathers' slaveholdings.

1820: no listing, Va. census.
1830: 3 sls., Richmond, Va.
1840: 0 sls., Brazoria, Co., Tex.
1850: 0 sls., Brazoria, Co., Tex.

3. WILLIAM H. BULLOCK, SR. (?-1829), planter
1790: Granville Co., N.C.; no slaveholdings listed for this county.
1800: 0 sls., Granville Co., N.C.
1810: 59 sls., Granville Co., N.C.
1820: 82 sls., Granville Co., N.C.
1830: 24 sls., Granville Co., N.C.
Four daughters, four sons:

Richard Bullock (c. 1774-?), planter†
1800: 55 sls., Warren Co., N.C.
1810: 59 sls., Warren Co., N.C.
1820: 97 sls., Warren Co., N.C.
1830: 18 sls., Granville Co., N.C.
1840: 125 sls., Warren Co., N.C.
1850: 136 sls., Warren Co., N.C.
1860: 111 sls., Warren Co., N.C.

John Bullock (c. 1796-?), planter
1820: no listing, N.C. census.
1830: 22 sls., Granville Co., N.C.
1840: 34 sls., Granville Co., N.C.
1850: 67 sls., Granville Co., N.C.
1860: 67 sls., Granville Co., N.C.

*William H. Bullock (c. 1803-?), planter
Migrated from North Carolina to Alabama c. 1830.
1830: 31 sls., Caswell Co., N.C.
1840: 23 sls., Greene Co., Ala.
1850: 48 sls., Greene Co., Ala.
1860: 36 sls., Greene Co., Ala.

James M. Bullock (c. 1810-?), planter†
1830: 0 sls., Granville Co., N.C.
1840: 54 sls., Granville Co., N.C.
1850: 22 sls., Granville Co., N.C.
1860: 133 sls., Granville Co., N.C.

4. LEWIS BURWELL (1764-1847), planter
1790: census does not list slaveholdings.
1800: census destroyed.
1810: 9 sls., Franklin Co., Va.
1820: 48 sls., Mecklenburg Co., Va.

*Migrants
†Sons who equaled or surpassed their fathers' slaveholdings.

1830: 3 sls., Richmond, Va.
1840: no listing, Va census.
1847: 21 slaves mentioned in his will.
One daughter, five sons:

*Abraham Burwell (c. 1805-?), planter and mechanic
Date of migration from Virginia to Mississippi unknown.
1830: no listing, Va. or Miss. censuses.
1840: no listing, Va. or Miss. censuses.
1850: 0 sls., Lauderdale Co., Miss.
1860: 20 sls., Lauderdale Co., Miss.

Allen Burwell (c. 1806-?), farmer
1830: no listing, Va. census.
1840: no listing, Va. census.
1850: 11 sls., Lunenburg Co., Va.
1860: 13 sls., Lunenburg Co., Va.

William Armistead Burwell (c. 1810-died during the Civil War), planter
1830: 4 sls., Dinwiddie Co., Va.
1840: 6 sls., Dinwiddie Co., Va.
1850: 26 sls., Franklin Co., Va.
1860: no listing, Franklin Co., Va.

*Lewis Burwell (c. 1811-?), planter
Migrated from Virginia to Alabama between 1840 and 1845.
1840: no listing, Va. or Ala. censuses.
1850: 14 sls., Marengo Co., Ala.
1860: 25 sls., Marengo Co., Ala.

John R. Burwell (c. 1814-?), farmer
1840: no listing, Va. census.
1850: 8 sls., Franklin Co., Va.
1860: no listing, Franklin Co., Va.

5. JAMES CALHOUN (1779-1843), planter
1800: 12 sls., Abbeville Co., S.C.
1810: no listing, S.C. census.
1820: 28 sls., Abbeville Co., S.C.
1830: 28 sls., Abbeville Co., S.C.
1840: no listing, S.C. census.
Five daughters, three sons:

*James M. Calhoun (1805-1877), planter and lawyer†
Migrated from South Carolina to Alabama c. 1831.
1830: no listing, S.C. or Ala. censuses.
1840: 41 sls., Dallas Co., Ala., and 13 sls., Perry Co., Ala.

*Migrants
†Sons who equaled or surpassed their fathers' slaveholdings.

1850: no listing, Ala. census.
1860: 168 sls., Dallas Co., Ala.

John A. Calhoun (c. 1807-?), planter†
1830: no listing, S.C. census.
1840: 24 sls., Abbeville Co., S.C.
1850: 103 sls., Abbeville Co., S.C.
1860: 135 sls., Abbeville Co., S.C.

William H. Calhoun (c. 1810s-?), farmer
1840: 8 sls., Abbeville Co., S.C.
1850: 6 sls., Abbeville Co., S.C.
1860: 13 sls., Abbeville Co., S.C.

6. JOHN C. CALHOUN (1782-1850), planter, lawyer, and politician
1810: 4 sls., Abbeville Co., S.C.
1820: 2 sls., Laurens Co., S.C.
1830: 34 sls., Abbeville Co., S.C.; a biographer says he owned a total of "some eighty" slaves.
1840: 69 sls., Pickens Co., S.C.
1850: 76 sls., owned by his widow, Floride C. Calhoun, Pickens Co., S.C.
Two daughters, five sons:

*Andrew P. Calhoun (1811-1865), planter†
Migrated from South Carolina to Alabama in 1838; returned to South Carolina c. 1860.
1840: 77 sls., Marengo Co., Ala.
1850: 115 sls., Marengo Co., Ala.
1860: 100 sls., Marengo Co., Ala.

Patrick Calhoun (1821-1858), soldier
1850: 0 sls., Abbeville Co., S.C.

*John C. Calhoun (1823-1855), farmer
Migrated from South Carolina to Florida c. 1850.
1850: 0 sls., Putnam Co., Fla., although he brought a "small" slave force with him to Florida; died a bankrupt.

*James E. Calhoun (1826-1861), lawyer
Migrated from South Carolina to California in early 1850s; died a bankrupt.

William L. Calhoun (1829-1858), farmer
1850: 6 sls., Abbeville Co., S.C.; died a bankrupt.

7. MICAJAH CARR (1752-1812), planter
1785: Resided in Albemarle Co., Va. State census did not list slaves.
1800: census destroyed.
1810: 20 sls., Albemarle Co., Va.
Five daughters, five sons:

*Migrants
†Sons who equaled or surpassed their fathers' slaveholdings.

James O. Carr (c. 1780-?), planter†
1810: 7 sls., Albemarle Co., Va.
1820: no listing, Va. census.
1830: 17 sls., Albemarle Co., Va.
1840: 18 sls., Albemarle Co., Va.
1850: 22 sls., Albemarle Co., Va.
1860: no listing, Albemarle Co., Va.

David Carr (c. 1785-?), planter†
1810: 0 sls., Isle of Wight Co., Va.
1820: 0 sls., Isle of Wight Co., Va.
1830: 27 sls., Albemarle Co., Va.
1840: 29 sls., Albemarle Co., Va.
1850: 43 sls., Albemarle Co., Va.
1860: no listing, Albemarle Co., Va.

*Burton Carr (c. 1792-?), whiskey distiller
Migrated from Virginia to Tennessee in the 1810s.
1820: living somewhere in Tenn., no listing in census.
1830: living somewhere in Tenn., no listing in census.
1840: living somewhere in Ky., no listing in census.
1850: 0 sls., Barren Co., Ky.
1860: no listing, Barren Co., Ky.

George Carr (c. 1802-?), planter and lawyer†
1830: 0 sls., Lee Co., Va.
1840: 5 sls., King and Queen Co., Va.
1850: no listing, Va. census.
1860: 21 sls., Albemarle Co., Va.

Name, birthdate, occupation, and date of death of fifth son are unknown.

8. JOHN HERBERT DENT (1782-1823), planter and sea captain
1810: 0 sls., Charleston, Co., S.C.
1820: 112 sls., Colleton Co., S.C.
Five daughters, two sons:

*John Horry Dent (1815-1892), planter
Migrated from South Carolina to Alabama in 1837.
1840: 55 sls., Barbour Co., Ala.
1850: 83 sls., Barbour Co., Ala.
1860: 100 sls., Barbour Co., Ala.

George C. Dent (1822-1884), planter†
Settled with his mother on her plantation in Darien, Georgia.
1850: 125 sls., McIntosh Co., Ga.
1860: no listing, McIntosh Co., Ga.

*Migrants
†Sons who equaled or surpassed their fathers' slaveholdings.

9. ISAIAH DUBOSE (1781–1857), planter
1810: 4 sls., Darlington Co., S.C.
1820: 50 sls., Darlington Co., S.C.
1830: 85 sls., Charleston Co., S.C.
1840: no listing, S.C. census.
1850: no listing, S.C. census.
Six daughters, three sons:

*Kimbrough C. DuBose (1809–?), planter†
Migrated from South Carolina to Alabama in 1850.
1830: 14 sls., Darlington Co., S.C.
1840: no listing, S.C. census.
1850: 151 sls., Marengo Co., Ala.
1860: 129 sls., Marengo Co., Ala.

*James H. DuBose (1811–?), planter†
Migrated from South Carolina to Alabama c. 1841.
1840: 0 sls., Marengo Co., Ala.
1850: 140 sls., Marengo Co., Ala.
1860: 166 sls., Marengo Co., Ala.

*Charles DuBose (1812–1850s), planter
Migrated from South Carolina to Alabama in the 1830s; returned to South Carolina before 1850.
1840: 9 sls., Marengo Co., Ala.
1850: 50 sls., Charleston Co., S.C.

10. JOHN FINLEY (1778–1865), planter and merchant
1800: 0 sls., Guilford Co., N.C.
1810: 21 sls., Wilkes Co., N.C.
1820: no listing, Wilkes Co., N.C.
1830: 27 sls., Wilkes Co., N.C.
1840: 19 sls., Wilkes Co., N.C.
1850: 20 sls., Wilkes Co., N.C.
1860: 22 sls., Wilkes Co., N.C.
One daughter, three sons:

*Augustus W. Finley (1812–1889), farmer
Migrated from North Carolina to Alabama in early 1850s; returned to Wilkes Co., N.C., before 1860.
1840: no listing, Wilkes Co., N.C.
1850: 13 sls., Wilkes Co., N.C.
1860: 17 sls., Wilkes Co., N.C.

*John T. Finley (1817–1896), farmer and merchant
Migrated from North Carolina to Alabama in 1847; returned to Wilkes Co., N.C., in late 1860 or in 1861.

*Migrants
†Sons who equaled or surpassed their fathers' slaveholdings.

1840: no listing, N.C. census.
1850: no listing in Ala. census; 8 sls. listed under his name in Wilkes Co., N.C.
1860: living in Cherokee Co., Ala., but no listing in the slave schedule; the slaves may have been en route to Wilkes Co., N.C.

William P. Finley (1849–1910). Lived with his parents in Wilkes Co., N.C.

11. JAMES HACKETT (c. 1780–1845), planter
1800: no listing, N.C. census.
1810: 11 sls., Wilkes Co., N.C.
1820: 0 sls., Guilford Co., N.C.
1830: 20 sls., Wilkes Co., N.C.
1840: 19 sls., Wilkes Co., N.C.
One daughter, five sons:

Alexander Hackett (c. 1821–?), farmer
1850: 4 sls., Wilkes Co., N.C.
1860: 9 sls., Wilkes Co., N.C.

Robert F. Hackett (1823–after 1882), physician
1850: 1 slave, Wilkes Co., N.C.
1860: 7 sls., Wilkes Co., N.C.

James W. Hackett (1824–?), farmer
1850: 1 slave, Wilkes Co., N.C.
1860: 10 sls., Wilkes Co., N.C.

*Richard R. Hackett (1826–c. 1863), lawyer
Migrated from North Carolina to Louisiana in 1848.
1850: 2 sls., Caddo Parish, La.
1860: no listing in Caddo Parish, La.

Charles Hackett. Youngest son; no information on his birthdate, occupation, or date of death. Evidently lived with one of his brothers in Wilkes Co., N.C., but is not listed in any household.
1860: 1 slave, Wilkes Co., N.C.

12. WILLIAM B. IRBY (1800–1896), planter
1820: 59 sls., Nottoway Co., Va.
1830: no listing, Nottoway Co., Va.
1840: 47 sls., Nottoway Co., Va.
1850: 45 sls., Nottoway Co., Va.
One daughter, three sons:

*Edmund Irby (c. 1810–?), planter
Migrated from Virginia to Mississippi in 1840s.
1830: no listing, Va. census.
1840: no listing, Va. census.

*Migrants

1850: 15 sls., Panola Co., Miss.
1860: 31 sls., Panola Co., Miss.

*John W. Irby (c. 1824–?), farmer and physician
Migrated from Virginia to Mississippi in 1848.
1850: 19 sls., Panola Co., Miss.
1860: 13 sls., Panola Co., Miss.

*Freeman Irby (c. 1828–?), planter
Migrated from Virginia to Mississippi in late 1840s.
1850: 57 sls., Panola Co., Miss.
1860: 36 sls., Panola Co., Miss.

13. WILLIAM LEA (c. 1776–1873), planter, merchant, and minister
1800: 4 sls., Person Co., N.C.
1810: no listing, N.C. census.
1820: 33 sls., Caswell Co., N.C.
1830: 63 sls., Caswell Co., N.C.
1840: 60 sls., Caswell Co., N.C.
1850: 24 sls., Caswell Co., N.C.
1860: 23 sls., Caswell Co., N.C.
One daughter, five sons:

*Willis M. Lea (1801–1879), planter and physician†
Migrated from North Carolina to Mississippi c. 1841.
1830: no listing, N.C. census.
1840: no listing, N.C. census.
1850: 72 sls., Marshall Co., Miss.
1860: 47 sls., Marshall Co., Miss.

*Lorenzo Lea (1806–1872), planter and minister
Migrated from North Carolina to Tennessee c. 1847.
1830: no listing, N.C. census.
1840: 0 sls., Caswell Co., N.C.
1850: 5 sls., Madison Co., Tenn.
1860: 43 sls., Madison Co., Tenn.

Solomon Lea (1807–1897), farmer, minister, and educator
1830: no listing, N.C. census.
1840: no listing, N.C. census.
1850: 8 sls., Caswell Co., N.C.
1860: 15 sls., Caswell Co., N.C.

*William Lea (c. 1809–1857), farmer and merchant
Migrated from North Carolina to Tennessee in 1853.
1840: 11 sls., Caswell Co., N.C.
1850: 17 sls., Caswell Co., N.C.
1860: His widow, Mary Wilson Lea, owned no slaves in Madison Co., Tenn

*Migrants
†Sons who equaled or surpassed their fathers' slaveholdings.

*Addison Lea (c. 1810–1855), farmer and minister
Migrated from North Carolina to Mississippi in 1852.
1830: no listing, N.C. census.
1840: no listing, N.C. census.
1850: 10 sls., Caswell Co., N.C.

14. ABRAHAM NOTT (1768–1830), planter, politician, and jurist
1790: no listing, S.C. census.
1800: 7 sls., Union Co., S.C.
1810: 12 sls., Richland Co., S.C.
1820: 11 sls., Richland Co., S.C.
1830: His widow, Angelica Nott, owned 65 slaves in Richland Co., S.C.
Four daughters, six sons:

William B. Nott, oldest son (1790s–?), physician
1820: no listing, S.C. census.
1830: 14 sls., Union Co., S.C.
1840: no listing, S.C. census.
1850: no listing, S.C. census.
1860: no listing, S.C. census.

Henry J. Nott (1797–1837), lawyer and writer
1820: no listing, S.C. census.
1830: 3 sls., Richland Co., S.C.

*Rufus Nott (c. 1803–?), planter and physician
Migrated from South Carolina to Louisiana in the 1840s.
1830: no listing, S.C. census.
1840: 41 sls., Richland Co., S.C.
1850: 30 sls., Caddo Parish, La.
1860: 0 sls., Refugio Co., Tex.

*Josiah Nott (1804–1873), physician and proslavery writer
Migrated from South Carolina to Alabama in 1835.
1830: no listing, S.C. census.
1840: no listing, Ala. census.
1850: 16 sls., Mobile, Ala.
1860: 10 sls., Mobile, Ala.

*James Nott (c. 1810s–?), physician
Migrated from South Carolina to Louisiana in the 1830s.
1830: no listing, S.C. census.
1840: 0 sls., New Orleans, La.
1850: 0 sls., Victoria Co., Tex.
1860: 0 sls., Victoria Co., Tex.

*Gustavus Nott (?–?), physician
Migrated from South Carolina to Louisiana in the 1830s.

*Migrants

1830: no listing, S.C. census.
1840: 0 sls., New Orleans, La.
1850: no listing, New Orleans census.

15. ISAAC OTEY (1767–1855), planter and politician
1790: census does not list slaves.
1800: census destroyed.
1810: no listing, Va. census.
1820: 10 sls., Bedford, Co., Va.
1830: 33 sls., Bedford, Co., Va.
1840: 14 sls., Bedford, Co., Va.
1850: 15 sls., Bedford, Co., Va.
Four daughters, eight sons:

Isaac N. Otey (1790s–?), planter†
1810: 4 sls., New Kent Co., Va.
1820: 7 sls., Bedford, Co., Va.
1830: 33 sls., Bedford, Co., Va.
1840: 14 sls., New Kent Co., Va.
1850: 15 sls., New Kent Co., Va.
1860: 0 sls., Bedford, Co., Va.

John M. Otey (1792–1859), planter†
1820: 2 sls., Bedford, Co., Va.
1830: 12 sls., Campbell Co., Va.
1840: 35 sls., Campbell Co., Va.
1850: 12 sls., Campbell Co., Va.

Robert Otey (1790s–?)
No listing in Va. censuses; evidently died as a youth.

Littleton Otey (1790s–?)
No listing in Va. censuses; evidently died as a youth.

*Armistead G. Otey (1797–1863), farmer
Migrated from Virginia to Mississippi between 1850 and 1860.
1830: 9 sls., Bedford, Co., Va.
1840: 16 sls., Bedford, Co., Va.
1850: no listing, Va. or Miss. censuses.
1860: 10 sls., Holmes Co., Miss.

*James H. Otey (1800–1863), minister
Migrated from Virginia to Tennessee c. 1827.
1820: no listing, Va. census.
1830: 3 sls., Williamson Co., Tenn.
1840: 4 sls., Maury Co., Tenn.
1850: 7 sls., Maury Co., Tenn.
1860: no listing, Maury Co., Tenn.

*Migrants
†Sons who equaled or surpassed their fathers' slaveholdings.

*Walter L. Otey (1806–1876), planter†
Migrated from Virginia to Arkansas, date unknown.
1830: no listing, Va. census.
1840: no listing, Va. or Ark. censuses.
1850: no listing, Va. or Ark. censuses.
1860: 43 sls., Phillips Co., Ark.

William L. Otey (c. 1811–?), farmer
1830: 3 sls., Charles City Co., Va.
1840: 0 sls., Bedford, Co., Va.
1850: 3 sls., Charles City Co., Va.
1860: 5 sls., Charles City Co., Va.

16. SAMUEL PICKENS (1743–1821), planter
1790: 5 sls., Mecklenburg Co., N.C.
1800: 9 sls., Cabarrus Co., N.C.
1810: 19 sls., Cabarrus Co., N.C.
1820: 20 sls., Cabarrus Co., N.C.
Two daughters, five sons:

James Pickens (c. 1775–c. 1817), planter†
1800: no listing, N.C. census.
1820: 0 sls., Guilford Co., N.C.
Bequeathed 61 slaves to a nephew in 1817.

*John Pickens (c. 1770s–1850s), farmer
Migrated from North Carolina to Alabama in the 1820s.
1800: no listing, N.C. census.
1810: 0 sls., Guilford Co., N.C.
1820: no listing in N.C. or Ala. censuses.
1830: 0 sls., Clarke Co., Ala.
1840: 50 sls., Greene Co., Ala.
1850: no listing, Greene Co., Ala.

*Israel Pickens (1780–1827), lawyer, planter, and politician
Migrated from North Carolina to Alabama in 1817.
1810: 3 sls., Mecklenburg Co., N.C.
1820: no listing in Ala. census.

*Samuel Pickens (1782–1855), planter and politician†
Migrated from North Carolina to Alabama c. 1817.
1810: 0 sls., Mecklenburg Co., N.C.
1820: census destroyed.
1830: 25 sls., Greene Co., Ala.
1840: 62 sls., Greene Co., Ala.
1850: 124 sls., Greene Co., Ala.

*Migrants
†Sons who equaled or surpassed their fathers' slaveholdings.

Robert Pickens (?-?)
No listing in N.C. censuses; probably died as a youth, but birthdate, occupation, and date of death are unknown.

17. WILLIAM POLK (1758-1834), planter, banker, and politician
1790: 2 sls., Mecklenburg Co., N.C.
1800: 9 sls., Wake Co., N.C.; Tenn. census destroyed.
1810: no listing, N.C. census; Tenn. census destroyed.
1820: 31 sls., Mecklenburg Co., N.C.; 12 sls., Maury Co., Tenn.
1830: 41 sls., Wake Co., N.C.; 97 sls., Maury Co., Tenn.
Six daughters, nine sons:

*Thomas G. Polk (c. 1787-?), planter and politician
Migrated from North Carolina to Tennessee in the 1830s.
1810: 6 sls., Anson Co., N.C.
1820: 44 sls., Mecklenburg Co., N.C.
1830: 33 sls., Mecklenburg Co., N.C.
1840: 72 sls., Fayette Co., Tenn.
1850: 73 sls., Fayette Co., Tenn.
1860: no listing, Tenn. census.

*William J. Polk (1790s-?), planter and physician
Migrated from North Carolina to Tennessee c. 1836.
1810: 0 sls., Mecklenburg Co., N.C.
1820: 83 sls., Mecklenburg Co., N.C.
1830: 19 sls., Wake Co., N.C.
1840: 13 sls., Maury Co., Tenn.
1850: 64 sls., Maury Co., Tenn.
1860: no listing, Maury Co., Tenn.

*Lucius J. Polk (c. 1802-?), planter
Migrated from North Carolina to Tennessee in 1823.
1830: 51 sls., Maury Co., Tenn.
1840: 52 sls., Maury Co., Tenn.
1850: 61 sls., Maury Co., Tenn.
1860: 81 sls., Maury Co., Tenn.

*Leonidas Polk (1806-1864), planter and minister†
Migrated from North Carolina to Tennessee in 1833; migrated from Tennessee to Louisiana in 1841; migrated from Louisiana to Tennessee in 1860.
1830: Studying in a Virginia seminary. No listing in N.C. or Tenn. censuses.
1840: 105 sls., Maury Co., Tenn.
1850: 216 sls., Lafourche Parish, La.
1860: no listing, La. or Tenn. censuses; en route from Louisiana to Sewanee, Tenn., in the summer of 1860.

*Migrants
†Sons who equaled or surpassed their fathers' slaveholdings.

*Rufus K. Polk (c. 1815–1843), planter
Migrated from North Carolina to Tennessee in 1835.
1840: no listing, Maury Co., Tenn.

*George W. Polk (c. 1817–?), planter
Migrated from North Carolina to Tennessee c. 1840.
1840: 5 sls., Obion Co., Tenn.
1850: 67 sls., Maury Co., Tenn.
1860: 82 sls., Maury Co., Tenn.

Alexander Hamilton Polk (?–1830)
Died at his father's home when he was in his late teens or early twenties.
1820: no listing, N.C. census.
1830: no listing, N.C. census.

Charles J. Polk (?–1830)
Died as a child.

*Andrew J. Polk (1828–?), planter†
Migrated from North Carolina to Tennessee c. 1847.
1850: 168 sls., Maury Co., Tenn.
1860: 96 sls., Maury Co., Tenn.

18. JOHN TAYLOE III (1771–1828), planter and iron manufacturer
1790: 65 sls., Essex Co., Va.; 180 sls., Richmond Co., Va.
1800: census destroyed.
1810: 94 sls., King George Co., Va.
1820: 72 sls., King George Co., Va.
1828: owned between 750 and 780 slaves at his death.
Seven daughters, seven sons:

John Tayloe IV (?–1824), planter
1820: 1 slave, Essex Co., Va.

Benjamin Ogle Tayloe (1796–1868), planter
1820: no listing, Va. census.
1830: 78 sls., Prince George Co., Va.; 72 sls., King William Co., Va.
1840: no listing, Va. census.
1850: no listing, Va. census; 107 sls., Marengo Co., Ala. (Managed by his brother Henry.)
1860: 149 sls., Marengo Co., Ala. (Managed by his brother Henry.)

William H. Tayloe (1799–1871), planter
1820: no listing, Va. census.
1830: no listing, Va. census.
1840: 0 sls., King George Co., Va.
1850: no listing, Va. census; 71 sls., Marengo Co., Ala., under his name; 62 sls., Marengo Co., under the name of "W. H. Tayloe and Company."

*Migrants
†Sons who equaled or surpassed their fathers' slaveholdings.

1860: 152 sls., Perry Co., Ala.; 125 sls., Marengo Co., Ala.; no listing, Va. census, although he resided in Va.

Edward T. Tayloe (1803–1876), planter and diplomat
1840: 76 sls., King George Co., Va.
1850: 94 sls., King George Co., Va.
1860: 85 sls., King George Co., Va., and 104 sls., Perry Co., Ala.

George P. Tayloe (1804–1897), planter
1830: 8 sls., Essex Co., Va.
1840: no listing, Va. census.
1850: 59 sls., Marengo Co., Ala., under his name; 26 sls., Marengo Co., under name of "G. P. Tayloe trustee"; 70 sls., Marengo Co., under name of "G. P. Tayloe and Company"; no listing, Va. census, although he resided in Va.
1860: 67 sls., Marengo Co., Ala.

*Henry A. Tayloe (1808–1903), planter
Migrated from Virginia to Alabama in 1834.
1830: 69 sls., Essex Co., Va.
1840: 302 sls., Marengo Co., Ala.
1850: no listing, Ala. census.
1860: 54 sls., Marengo Co., Ala.; 168 sls., Richmond Co., Va.

Charles Tayloe (1810–1847), diplomat
1830: no listing, Va. census.
1840: no listing, Va. census.
1850: 18 sls., estate of Charles Tayloe, King George Co., Va.

19. SAMUEL TOWNES (1773–1826), planter
1800: 3 sls., Greenville Co., S.C.
1810: no listing, S.C. census.
1820: 25 sls., Greenville Co., S.C.
1827: 27 sls., his estate, Greenville Co., S.C.
One daughter, five sons:

Henry Townes (1804–1849), farmer and physician
1830: 5 sls., Greenville Co., S.C.
1840: 15 sls., Abbeville Co., S.C.

*Samuel A. Townes (1806–1893), farmer, lawyer, and journalist
Migrated from South Carolina to Alabama in 1834; returned to South Carolina in late 1840s.
1830: no listing, S.C. census.
1840: 10 sls., Perry Co., Ala.
1850: 8 sls., Greenville Co., S.C.
1860: 13 sls., Greenville Co., S.C.

*Migrants

George F. Townes (1809–1891), farmer, lawyer, journalist, and politician
1830: no listing, S.C. census.
1840: 3 sls., Greenville Co., S.C.
1850: 10 sls., Greenville Co., S.C.
1860: 17 sls., Greenville Co., S.C.

John A. Townes (1812–1862), planter
1840: no listing, S.C. census.
1850: lived with his mother, Rachel Townes, who owned 35 slaves in Greenville Co., S.C.
1860: 25 sls., Greenville, Co., S.C.

William A. Townes (1821–1873), planter
1850: lived with his mother, Rachel Townes, Greenville Co., S.C.
1860: 25 sls., Greenville Co., S.C.

20. NEEDHAM WHITFIELD (1758–1812), planter
1790: 27 sls., Wayne Co., N.C.
1800: 60 sls., Wayne Co., N.C.
1810: 62 sls., Wayne Co., N.C.
Six daughters, five sons:

*William Whitfield (1783–1864), planter†
Migrated from North Carolina to Mississippi in 1829.
1810: no listing, N.C. census.
1820: 0 sls., Duplin Co., N.C.
1830: no listing, Lowndes Co., Miss.
1840: no listing, Lowndes Co., Miss.
1850: 12 sls., Lowndes Co., Miss.
1860: 69 sls., Lowndes Co., Miss.

*Needham Whitfield (1789–after 1871), planter and physician†
Migrated from North Carolina to Mississippi between 1830 and 1837.
1810: no listing, N.C. census.
1820: no listing, N.C. census.
1830: 64 sls., Lenoir Co., N.C.
1840: 77 sls., Monroe Co., Miss.
1850: 94 sls., Monroe Co., Miss.
1860: 81 sls., Monroe Co., Miss.

*Edmund Whitfield (1793–1867), planter
Migrated from North Carolina to Mississippi in 1840.
1820: 13 sls., Wayne Co., N.C.
1830: 20 sls., Wayne Co., N.C.
1840: 25 sls., Monroe Co., Miss.
1850: 43 sls., Monroe Co., Miss.
1860: 40 sls., Monroe Co., Miss.

*Migrants
†Sons who equaled or surpassed their fathers' slaveholdings.

*Gaius Whitfield (1804–after 1865), planter†
Migrated from North Carolina to Alabama c. 1830.
1830: 38 sls., Marengo Co., Ala.
1840: 128 sls., Marengo Co., Ala.
1850: 350 sls., Marengo Co., Ala.
1860: 383 sls., Marengo Co., Ala., and 83 sls., Lowndes Co., Miss.

*Boaz Whitfield (1806–1843), physician
Migrated from North Carolina to Alabama between 1832 and 1840.
1830: no listing, N.C. census.
1840: no listing, Ala. census.

---

*Migrants
†Sons who equaled or surpassed their fathers' slaveholdings.

Table 3. Sex Ratios among Slaves, Selected Seaboard Counties, 1820, 1830, 1840

County	1820	1830	1840
Charlotte Co., Va.	110.7	107.6	105.6
	(2,080)	(5,631)	(5,614)
Powhatan Co., Va.	112.2	118.0	114.6
	(1,114)	(3,329)	(3,105)
Bertie Co., N.C.	77.9	107.9	101.8
	(1,779)	(4,004)	(3,964)
Nash Co., N.C.	109.1	97.9	93.3
	(1,836)	(2,033)	(2,022)
Darlington Co., S.C.	106.9	95.8	95.4
	(2,512)	(4,225)	(4,446)
Sumter Co., S.C.	125.2	97.5	93.3
	(3,522)	(10,889)	(11,326)
Total males	6,675	15,249	15,154
Total females	6,168	14,862	15,323
Total slaves	12,843	30,111	30,477

Source: See "A Note on the Tables." The ratio is the number of male slaves for every 100 female slaves. Parentheses contain the number of male and female slaves between the ages of ten and fifty-five in each county.

Table 4. Sex Ratios among Slaves, Selected Southwestern Counties, 1830, 1840, 1850, 1860

County	1830	1840	1850	1860
Marengo Co., Ala.	92.5	108.0	107.5	101.4
	(1,841)	(7,905)	(12,841)	(13,166)
Perry Co., Ala.	102.6	116.9	108.6	106.8
	(2,652)	(6,879)	(8,699)	(10,106)
DeSoto Co., Miss.[1]	—	104.6	108.0	100.7
	—	(1,886)	(6,041)	(8,888)
Hinds Co., Miss.	104.1	107.2	109.4	100.4
	(1,941)	(8,153)	(10,255)	(14,294)
Guadalupe Co., Tex.	—	—	95.4	94.3
	—	—	(215)	(1,059)
Harrison Co., Tex.	—	—	93.2	103.8
	—	—	(3,926)	(5,536)
Total males	3,218	12,995	21,675	26,823
Total females	3,216	11,828	20,302	26,226
Total slaves	6,434	24,823	41,977	53,049

Source: See "A Note on the Tables." The ratio is the number of male slaves for every 100 female slaves. Parentheses contain the number of male and female slaves between the ages of ten and fifty-five in each county.

1. Formed in 1836.

Table 5. Household Structure among Planter Families, Selected Southwestern Counties, 1840, 1850, 1860

County	N	Nuclear	Complex	Ambiguous
Marengo Co., Ala.	64	65%	17%	17%
Perry Co., Ala.	66	53	20	27
DeSoto Co., Miss.	42	50	28	21
Hinds Co., Miss.	83	44	25	30
Guadalupe Co., Tex.	7	43	57	0
Harrison Co., Tex.	28	57	14	28
Total N	290			
Percentage of total		53	23	24

Source: See "A Note on the Tables." N = number of households.

Notes

The following abbreviations are used in the notes for the location of manuscript collections:

ALA	Alabama Department of Archives and History
DU	Perkins Library, Duke University
LC	Library of Congress
MS	Mississippi Department of Archives and History
NCA	North Carolina Department of Archives and History
SCA	South Carolina Department of Archives and History
SCHS	South Carolina Historical Society
UNC	Southern Historical Collection, University of North Carolina
USC	South Caroliniana Library, University of South Carolina
UT	Eugene C. Barker Texas History Center, University of Texas at Austin
UVA	Alderman Library, University of Virginia
VHS	Virginia Historical Society
VSA	Virginia State Archives

Introduction

1. This book focuses for the most part on men who went west seeking change, so it tells one aspect of the story of the westward movement.

2. Here I can only summarize the exciting, multifaceted literature on the planter family. Regarding the distribution of power within the family, Steven M. Stowe, *Intimacy and Power in the Old South: Ritual in the Lives of Planters*, New Studies in American Intellectual and Cultural History, Thomas Bender, consulting editor (Baltimore: Johns Hopkins University Press, 1987); Orville Vernon Burton, *In My Father's House Are Many Mansions: Family and Community in Edgefield, South Carolina*, Fred W. Morrison Series in Southern Studies (Chapel Hill: University of North Carolina Press, 1985); Bertram Wyatt-Brown, *Southern Honor: Ethics and*

Behavior in the Old South (New York: Oxford University Press, 1982); Michael P. Johnson, "Planters and Patriarchy: Charleston, 1800-1860," *Journal of Southern History* 66 (February 1980): 45-72, all portray the seaboard family as patriarchal. Jane Turner Censer, *North Carolina Planters and Their Children, 1800-1860* (Baton Rouge: Louisiana State University Press, 1984), portrays planter families of North Carolina as egalitarian, bourgeois, and on the whole similar to elite Northern families of the antebellum era. Burton implies that the affection family members felt for each other softened the worst aspects of the patriarchy, and Censer highlights the affection and harmony of family life.

Regarding the structure of the planter family, Burton and Censer both portray the family as nuclear; Stowe and Johnson do not discuss structure explicitly but assume that most families were nuclear. Daniel Blake Smith, *Inside the Great House: Planter Family Life in Eighteenth-Century Chesapeake Society* (Ithaca: Cornell University Press, 1980), argues that the planter family was transformed into a nuclear unit by the early nineteenth century. Wyatt-Brown, however, discusses the importance of relationships between extranuclear kinfolk, as do Robert C. Kenzer, *Kinship and Neighborhood in a Southern Community: Orange County, North Carolina, 1849-1881* (Knoxville: University of Tennessee Press, 1987), and Catherine Clinton, *The Plantation Mistress: Woman's World in the Old South* (New York: Pantheon Books, 1982), although none of these scholars explicitly addresses the issue of family structure.

Daniel Blake Smith, *Inside the Great House*, and Jan Lewis, *The Pursuit of Happiness: Family and Values in Jefferson's Virginia* (Cambridge; Cambridge University Press, 1983), are primarily concerned with affect, not structure. Both argue that planters became more affectionate with nuclear relatives than in the mid-eighteenth century and expected great happiness from these relationships; Smith explains this transformation by increasing affluence, secularization, literacy, and geographic mobility, and Lewis by the advent of evangelical religion combined with republicanism and domestic ideology.

3. Anne Firor Scott, *The Southern Lady: From Pedestal to Politics, 1830-1930* (Chicago: University of Chicago Press, 1970); Clinton, *Plantation Mistress*; Suzanne Lebsock, *The Free Women of Petersburg: Status and Culture in a Southern Town, 1784-1860* (New York: W. W. Norton & Company, 1984), xix, 115. Historians have debated extensively Mary Boykin Chesnut's comments on slavery in her famous diary. As I read the diary, her remarks support the general arguments of Scott, Clinton, and Lebsock, although other scholars suspect that Chesnut's antislavery views were insincere or believe that she exaggerated the extent of antislavery sentiment among other planter women. See Elizabeth Fox-Genovese, *Within the Plantation Household: Black and White Women of the Old South* (Chapel Hill: University of North Carolina Press, 1988), 339-65; *Mary Chesnut's Civil War*, ed. C. Vann Woodward (New Haven: Yale University Press, 1981), xlix.

4. Fox-Genovese, *Within the Plantation Household*, 35, 225; George C. Rable, *Civil Wars: Women and the Crisis of Southern Nationalism*, Women in American History Series, ed. Mari Jo Buhle, Jacquelyn D. Hall, and Anne Firor Scott (Urbana: University of Illinois Press, 1979), 37. Fox-Genovese also argues (64-65) that Southern society was paternalistic, not patriarchal. Jean E. Friedman, *The En-*

closed Garden: Women and Community in the Evangelical South, 1830–1900 (Chapel Hill: University of North Carolina Press, 1985), focuses on the role of the family and the church rather than class in women's lives; she argues that kinship and the church bound women to the family and to men and that women did not identify with each other as women.

5. See Eugene D. Genovese, *The Political Economy of Slavery: Studies in the Economy and Society of the Slave South* (New York; Pantheon Books, 1965), in which this view is most forcefully expressed, and modifications of it in later works: Eugene D. Genovese and Elizabeth Fox-Genovese, *Fruits of Merchant Capital: Slavery and Bourgeois Property in the Rise and Expansion of Capitalism* (New York: Oxford University Press, 1983); Eugene D. Genovese, *The World the Slaveholders Made: Two Essays in Interpretation* (New York: Pantheon Books, 1969); Eugene D. Genovese, *Roll, Jordan, Roll: The World the Slaves Made* (New York: Pantheon Books, 1974). James Oakes, *The Ruling Race: A History of American Slaveholders* (New York: Alfred A. Knopf, 1982), emphasizes the acquisitiveness of planter migrants on the Southwestern frontier; despite the many useful features of the book, he does not discuss the resistance to pure materialism among older planter men, and, while he notes that women experienced migration differently from men, he does not detail women's roles in the family.

Historians have overworked the concept of modernization to explain fundamental changes over time, but the theory still offers some insights into the deep changes that affected Americans in the eighteenth, nineteenth, and twentieth centuries. Modernization can be broadly defined as the transformation from a hierarchical, deferential society, in which kinship ties largely determine status, to a dynamic, open society in which status is a function of occupation, achievement, or some other ascriptive trait. This transformation also involves a change in outlook on the world, from one that is localistic, religious or mystical, and noncommercial to one that is cosmopolitan, rational, and commercial. Patterns of geographic mobility also change, from circulation in familiar routes near home to movement over vast distances. Finally, modernization involves a change in attitudes toward change itself: in a premodern society people value stability and see change as threatening, while innovation in all phases of life characterizes a modern society. See Richard D. Brown, *Modernization: The Transformation of American Life 1600–1865* (New York: Hill and Wang, 1976), 3–23.

Jack Temple Kirby conveys this last point—the fear of change in traditional societies—with great poignancy. He dates modernization as late as 1920 to 1960 for the masses of Southern white and black people; he defines modernization as the mechanization of agriculture and the end of economic self-sufficiency. He hints that men and women experienced this transformation differently; he notes, for instance, that men began using farm machinery before women (195). His chapter on women (Chapter 5) describes the hardships of women in the rural South but does not link them to theories of modernization. See *Rural Worlds Lost: The American South 1920–1960* (Baton Rouge: Louisiana State University Press, 1987), xiv–xv, 117, 369–70. For a discussion of how historians have employed the modernization concept, see the spring 1978 issue of *Social Science History*.

6. Wilbur J. Cash, *The Mind of the South* (New York: Alfred A. Knopf, 1941);

Frederick Jackson Turner, "The Significance of the Frontier in American History," in *The Turner Thesis Concerning the Role of the Frontier in American History*, 3rd ed., ed. and with an introduction by George Rogers Taylor (Lexington, Mass.: D. C. Heath, 1972), 28, 4. Turner did not include the experiences of women, of course, and he scarcely mentioned slavery. Recently historians have begun to discuss the destructiveness of the settlement process; see, for example, Patricia Nelson Limerick, *The Legacy of Conquest: The Unbroken Past of the American West* (New York: W. W. Norton & Company, 1987).

7. James D. Foust, *The Yeoman Farmer and Westward Expansion of United States Cotton Production* (New York: Arno Press, 1975), 70, 78, uses this definition for the Southwest. See "A Note on the Tables" for a discussion of these statistics.

Chapter 1. The Ties of Nature: The Planter Family in the Seaboard

1. Ellen Hard Townes, "Genealogy of the Townes Family," 1-9, William Townes to Samuel A. Townes, 26 August 1819, Will of Samuel Townes, written 3 September 1825, S. A. Townes to "My Dear Brother," 22 November 1826, S. A. Townes to his mother, 9 March 1829, S. A. Townes to "Dear Brother," 7 October 1829, Henry H. Townes to George Franklin Townes, 3 July 1831, H. H. Townes to G. F. Townes, 1 November 1832, H. H. Townes to John A. Townes, 19 May 1833, S. A. Townes to G. F. Townes, 24 October 1836, S. A. Townes to H. H. Townes, 17 August 1844, and other correspondence in the Townes Family Papers, USC; S. S. Crittenden, *The Greenville Century Book* (Greenville, S.C.: Greenville News, 1903), 25. Because several members of the Calhoun family married their cousins, individuals had nieces and nephews who were also their cousins. This explains why Armistead Burt is described as the husband of John Calhoun's cousin as well as the husband of his niece, who were one and the same person, Martha Calhoun Burt. See John Niven, *John C. Calhoun and the Price of Union*, Southern Biography Series, ed. William J. Cooper, Jr. (Baton Rouge: Louisiana State University Press, 1988), 339; *Papers of John C. Calhoun*, Volume 12, *1833-1835*, ed. Clyde N. Wilson (Columbia: published by University of South Carolina Press for South Carolina Department of Archives and History and South Caroliniana Society, 1979), 16; Lewis Perrin, "The Captains and the Kings Depart: Honorable Armistead Burt," *Press and Banner and Abbeville* (S.C.) *Medium*, 2 February 1933. Burt's later career as a congressman is described in the Armistead Burt Papers, DU.

2. Lawrence Stone, *The Family, Sex, and Marriage in England, 1500-1800* (New York: Harper & Row, 1977), 26; Edward Shorter, *The Making of the Modern Family* (New York: Basic Books, 1977), 205. See the introduction for a discussion of family structure.

3. This incidence of complex families is higher than those found for other American communities (between 5 and 25 percent) between 1800 and 1860; see Steven Ruggles, *Prolonged Connections: The Rise of the Extended Family in Nineteenth-Century England and America*, Social Demography Series, ed. Doris P. Slesinger, James A. Sweet, and Karl E. Taueber (Madison: University of Wisconsin Press, 1987), 5. Among scholars of the planter family, only Orville Vernon Burton, *In My Father's House Are Many Mansions: Family and Community in Edgefield,*

South Carolina, Fred W. Morrison Series in Southern Studies (Chapel Hill: University of North Carolina Press, 1985), 109–14, analyzes census returns for household membership. In the returns for 1850 and 1860, he finds that approximately half of wealthy households contained nuclear families; the rest contained other kin or lodgers. He also states (109–14) that wealthy households were larger than less affluent households but provides no numbers on household size.

4. The average number of slaves owned by each male head of household in Table 1 was fifty-three. The average household size is larger than the national average (based on the federal census returns) of five persons for the antebellum period. See Robert V. Wells, *Revolutions in Americans' Lives: A Demographic Perspective on the History of Americans, Their Families, and Their Society*, Contributions in Family Studies, No. 6 (Westport, Conn.: Greenwood Press, 1982), 150–52. James E. Davis, *Frontier America, 1800–1840: A Comparative Demographic Analysis of the Frontier Process* (Glendale, Calif.: Arthur H. Clark Company, 1977), 69, Table 1, finds an average household size of five persons for white families of all social classes in selected Southern counties, most of which are in the seaboard, for the censuses of 1810, 1820, and 1830.

5. A few scholars mention the visiting practices among planters, but no one has explored what these practices meant for the structure of the family. See Jane Turner Censer, *North Carolina Planters and Their Children, 1800–1860* (Baton Rouge: Louisiana State University Press, 1984), 8; Daniel Blake Smith, *Inside the Great House: Planter Family Life in Eighteenth-Century Chesapeake Society* (Ithaca: Cornell University Press, 1980), 77–79, 197; Bertram Wyatt-Brown, "The Ideal Typology and Antebellum Southern History: A Testing of a New Approach," *Societas: A Review of Social History* 5 (Winter 1975): 12; Paton Yoder, "Private Hospitality in the South, 1775–1850," *Mississippi Valley Historical Review* 47 (December 1960): 419–33.

6. Mrs. Burton Harrison [Constance Cary Harrison], *Recollections Grave and Gay* (New York: Charles Scribner's Sons, 1911), 22. See also Carrie and Addie Dogan to Caroline Gordon, 27 February 1854, Gordon and Hackett Family Papers, UNC; John Bullock to "My Dear Friend," 16 April 1855, John Bullock and Charles E. Hamilton Papers, UNC; Joe Gray Taylor, *Eating, Drinking, and Visiting in the South: An Informal History* (Baton Rouge: Louisiana State University Press, 1982), 62–63. Robert C. Kenzer documents the close proximity of the residences of kinfolk in antebellum Orange County, North Carolina, although he emphasizes the commitment of white Southerners to the neighborhoods in which they were born rather than kinship networks throughout the seaboard. See *Kinship and Neighborhood in a Southern Community: Orange County, North Carolina, 1849–1881* (Knoxville: University of Tennessee Press, 1987), 6–25.

7. Alicia H. Middleton to "Sister," 26 January 1840, Cheves-Middleton Papers, SCHS. On other ceremonial visits, see John J. Ambler, Jr., to Elizabeth B. Ambler, 5 June 1846, Ambler and Barbour Family Papers, UVA; J. G. H. Bullock to John Bullock, n.d. February 1853, John Bullock and Charles E. Hamilton Papers, UNC; Catherine Clinton, *The Plantation Mistress: Woman's World in the Old South* (New York: Pantheon Books, 1982), 67; Thomas Felix Hickerson, *Happy Valley: History and Genealogy* (Chapel Hill: by the author, 1940), 137.

8. Charles Ellis to Powhatan Ellis, 20 June 1830, Ellis Papers, UT; Louisa Cunningham to Benjamin C. Yancey, 8 April 1839, Benjamin Cudworth Yancey Papers, UNC; Elizabeth Blaetterman to "Mrs. Michie," n.d. [c. 1846], George Carr Papers, UVA.

9. H. H. Townes to Rachel Townes, 9 April 1844, Townes Family Papers, USC; Harrison *Recollections*, 22; Hickerson, *Happy Valley*, 59–63; Alicia Hopton Middleton, *Life in Carolina and New England During the Nineteenth Century* (Bristol, R.I.: by the author, 1929), 80.

10. *"Journal of a Secesh Lady": The Diary of Catherine Ann Devereux Edmonston, 1860–1866*, ed. Beth G. Crabtree and James W. Patton, 2nd printing (Raleigh: Division of Archives and History, Department of Cultural Resources, 1979), 614, 8 September 1864; Adam S. Dandridge to Mrs. R. M. T. Hunter, 8 November 1856, R. M. T. Hunter Papers, UVA. On activities during visits, see Andrew L. Pickens to Julia P. Howe, 3 September 1839, Israel Pickens Papers, ALA; [Junius?] A. Fox to James L. Gordon, 28 November 1841, Gordon and Hackett Family Papers, UNC; William Hunter to Thomas Hunter, n.d. 1849, Anna Gayle Fry Papers, ALA; John W. DuBose, "A Memoir of Four Families," 1:145, John Witherspoon DuBose Papers, ALA; Taylor, *Eating, Drinking, and Visiting*, 53–59; Thomas Felix Hickerson, *Echoes of Happy Valley: Letters and Diaries, Family Life in the South, Civil War History* (Chapel Hill: by the author, 1962), 32–33. On visiting springs, see Carrie and Addie Dogan to Caroline Gordon, 11 October 1848, Gordon and Hackett Family Papers, UNC; Hickerson, *Happy Valley*, 135; Censer, *Planters and Children*, 8–9; Wyatt-Brown, "Ideal Typology," 12.

11. Mira Lenoir to Julia Pickens, 7 October 1828, Chiliab Smith Howe Papers, UNC; Mary Ann Taylor to Marion Singleton, n.d. January 1834, Singleton Family Papers, LC; DuBose, "Memoir," 1:87, John Witherspoon DuBose Papers, ALA; Adolphus Blair to Mrs. Lewis E. Harvie, 12 September 1861, Harvie Family Papers, VHS. Like many Americans and Europeans, planters had begun to invest more emotion in all kinds of familial relationships, a trend that seems to have originated in the late seventeenth and early eighteenth centuries; see Peter N. Stearns and Carol Z. Stearns, "Emotionology: Clarifying the History of Emotions and Emotional Standards," *American Historical Review* 90 (October 1985): 818–21.

12. "Sallie" [Gwyn?] to Caroline Gordon, 29 August 1851, Gordon and Hackett Family Papers, UNC; Martha L. Finley to Caroline Gordon, 17 August 1846, James Gordon Hackett Papers, DU; Addie and Carrie Dogan to Caroline Gordon, 14 June 1848, Carrie Dogan to Caroline Gordon, 3 November 1856, Gordon and Hackett Family Papers, UNC.

13. Martha Pickens to Ann P. Jones, 21 October 1814, Israel Pickens Papers, ALA; Israel Pickens to William B. Lenoir, 1 January 1815, Chiliab Smith Howe Papers, UNC; *Papers of John C. Calhoun*, Volume 6, *1821–1822*, ed. W. Edwin Hemphill (Columbia: published by University of South Carolina Press for South Carolina Department of Archives and History and South Caroliniana Society, 1972), 392–94, John C. Calhoun to John Ewing Colhoun, 27 September 1821; *Calhoun Papers*, ed. Wilson, 13:641, Anna Maria Calhoun to Patrick Calhoun, 6 November 1834; Sarah D. Kennedy to Mrs. R. M. T. Hunter, 19 July 1846, R. M. T. Hunter Papers, UVA. Several historians note that colonial and antebellum women

could not travel alone but do not explore fully what this meant for family life; see Elizabeth Fox-Genovese, *Within the Plantation Household: Black and White Women of the Old South*, Gender and American Culture Series, ed. Linda K. Kerber and Nell Irvin Painter (Chapel Hill: University of North Carolina Press, 1988), 69; Clinton, *Plantation Mistress*, 176; Rhys Isaac, *The Transformation of Virginia 1740-1790* (Chapel Hill: published for Institute of Early American History and Culture, Williamsburg, Virginia, by University of North Carolina Press, 1982), 57.

14. John J. Ambler, Jr., to "My Dear Wife and Son," 26 February 1842, Ambler and Barbour Family Papers, UVA; Kate McLeod to Albert Blue, 25 March 1856, Matthew P. Blue Papers, ALA; C. W. Downing to "Aunt Sally" [probably Sarah C. Maddox], 22 January 1854, George Carr Papers, UVA.

15. Dabney Manuscript, UVA, 47–48. See also *Memorials of a Southern Planter: By Susan Dabney Smedes*, ed. and with an introduction and notes by Fletcher M. Green (New York: Alfred A. Knopf, 1965), 22.

16. Wells, *Revolutions*, 91–94.

17. Rita Jones Elliott, "The Herndon and Connor Families, Kith and Kin," unpublished paper, North Carolina State Library, Raleigh; correspondence in the Gordon and Hackett Family Papers, UNC; Hickerson, *Happy Valley*, map, n.p.; Mrs. James Harvey Gordon to Caroline Gordon and Sarah Gwyn Gordon Brown, 4 November 1854, Gordon and Hackett Family Papers, UNC; Hickerson, *Happy Valley*, 164, 197–203. David Hackett Fischer, *Albion's Seed: Four British Folkways in America* (New York: Oxford University Press, 1989), 646, notes that hundreds of members of the Calhoun family lived in the Carolina upcountry by the end of the eighteenth century.

18. William Henry Holcombe Autobiography and Diary, 21, UNC; Rebecca Sims to George C. Dromgoole, 22 February 1844, Edward Dromgoole Papers, UNC; J. Anderson to John Bullock, 19 July 1853, John Bullock and Charles E. Hamilton Papers, UNC. The private collection of this information preceded the founding of many genealogical societies in the late antebellum era.

19. M. E. Patterson to Neill Kelly, 14 March 1842, John N. Kelly Papers, DU; Zillah Haynie Brandon Diaries, 1:94, ALA; Holcombe Autobiography and Diary, 41, UNC.

20. David M. Schneider and Raymond T. Smith, *Class Differences and Sex Roles in American Kinship and Family Structure* (Englewood Cliffs, N.J.: Prentice-Hall, Inc., 1973), 16, note the difference between meaning and values in a culture and the ways that those values are expressed in specific actions.

21. Richard C. Ambler to "Dear Brother" [John J. Ambler], 2 February 1835, Ambler and Barbour Family Papers, UVA. Larry Schweikart, *Banking in the American South from the Age of Jackson to Reconstruction* (Baton Rouge: Louisiana State University Press, 1987), 54–57, skillfully describes some of the inadequacies of seaboard banks before the Crash of 1837, although he argues that seaboard banks were more stable than those of the Old Southwest. See Chapter 4 for a discussion of credit, banking, and kinship.

22. *Secret and Sacred: The Diaries of James Henry Hammond, a Southern Slaveholder*, ed. Carol Bleser (New York: Oxford University Press, 1988), 86–87, 170; Charles A. L. Lewis to R. M. T. Hunter, 29 November 1832, Thomas Hunter

to R. M. T. Hunter, 23 November 1832, R. M. T. Hunter Papers, UVA; Richard Morris to John Morris, 20 December 1846, John Morris Manuscripts, UVA; Diary of Martha T. W. Dyer, typescript, 12, 14 January 1823, Dyer-Plain Dealing Papers, UVA. The legal system allowed such reciprocal economic relationships, for Southern laws regarding insolvency, bankruptcy, and negotiability were lenient toward debtors. See Tony A. Freyer, "Law and the Antebellum Southern Economy: An Interpretation," in *Ambivalent Legacy: A Legal History of the South*, ed. David J. Bodenhamer and James W. Ely, Jr. (Jackson: University Press of Mississippi, 1984), 55-58. J. William Harris, *Plain Folk and Gentry in a Slave Society: White Liberty and Black Slavery in Augusta's Hinterlands* (Middletown, Conn.: Wesleyan University Press, 1985), 98-99, discusses economic exchanges among white men (some of whom were kin) in central Georgia.

23. Charles Cocke to Richard T. Archer, 30 November 1848, Archer Papers, UT; Elisha M. Carr to Mekins Carr, 20 February 1851, George Carr Papers, UVA; Ellis Malone to William Lea, 12 April 1838, Lea Family Papers, UNC.

24. Peter MacIntyre to Duncan McLaurin, 7 October 1837, Duncan McLaurin Papers, DU; C. W. Downing to "Aunt Sally" [probably Sarah C. Maddox], 22 January 1854, George Carr Papers, UVA. See also Robert McColley, *Slavery and Jeffersonian Virginia* (Urbana: University of Illinois Press, 1964), 32. The familial aspect of these exchanges distinguishes them from purely capitalistic market relationships and reflects the hybrid nature of the Southern economic system, which contained many premodern attributes despite the fact that it was geared toward producing commodities such as cotton and tobacco for the international market. For a thoughtful review of the debate on the nature of the early American economy, see Allan Kulikoff, "The Transition to Capitalism in Rural America," *William and Mary Quarterly*, 3rd ser., 46 (January 1989): 120-44.

25. On gifts and favors, see "Caroline" to Laura Lenoir and Julia Pickens, 27 December 1829, "Caroline" to Julia Pickens, 2 May 1830, Chiliab Smith Howe Papers, UNC; Elizabeth "F." to Fanny Hackett, 25 January 1830, Gordon and Hackett Family Papers, UNC; Hickerson, *Echoes of Happy Valley*, 19; *Calhoun Papers*, ed. Wilson, 17:429, Floride Calhoun to Margaret M. Calhoun, 10 September 1843. On housework, see *Growing Up in the 1850s: The Journal of Agnes Lee*, ed. Mary Custis Lee deButts (Chapel Hill: published for Robert E. Lee Memorial Association by University of North Carolina Press, 1984), 6-7; E. M. Hunter to Powhatan Ellis, 4 April 1820, Ellis Papers, UT; John J. Ambler, Jr., "Memoranda," 285, Ambler and Barbour Family Papers, UVA.

26. H. H. Townes to Rachel Townes, 26 October 1845, Lucretia A. Townes to Eliza T. Blassingame, 5 December 1832, Townes Family Papers, USC; Mira Lenoir to Julia Pickens, 13 February 1828, 9 July 1828, Chiliab Smith Howe Papers, UNC. Carol Gilligan, *In a Different Voice: Psychological Theory and Women's Development* (Cambridge, Mass.: Harvard University Press, 1982), illustrates the importance of relationships in the psychological makeup of contemporary women.

27. John J. Ambler, Jr., "Memoranda," passim, Ambler and Barbour Family Papers, UVA; Allan Kulikoff, "Historical Geographers and Social History: A Review Essay," *Historical Methods Newsletter* 6 (1973): 123; Wilbur Zelinsky, "The Hypothesis of the Mobility Transition," *Geographical Review* 61 (April 1971): 223.

Ronald S. Burt, *Toward a Structural Theory of Action: Network Models of Social Structure, Perception, and Action*, Quantitative Studies in Social Relations (New York: Academic Press, 1982), 4, notes that individuals perceive their interests differently and use social networks differently to achieve their goals. John T. Schlotterbeck, "The 'Social Economy' of an Upper South Community: Orange and Greene Counties, Virginia, 1815–1860," in *Class, Conflict, and Consensus: Antebellum Southern Community Studies*, Contributions in American History, No. 96 (Westport, Conn.: Greenwood Press, 1982), 19, argues that social networks were identical to community-based economic networks but does not distinguish between men's and women's social networks. Jean Friedman, *The Enclosed Garden: Women and Community in the Evangelical South, 1830–1900* (Chapel Hill: University of North Carolina Press, 1985), 3–20, argues that women's ties to their kinfolk and church members were more important than other social relationships and that relationships with other women per se were not very important.

28. "Caroline" to Julia Pickens, 28 August 1832, Amelia Douglass to Julia Pickens, 23 March 1833, Chiliab Smith Howe Papers, UNC; "Julia" to Caroline Gordon, 15 April 1847, Gordon and Hackett Family Papers, UNC; *Mary Chesnut's Civil War*, ed. C. Vann Woodward (New Haven: Yale University Press, 1981), xxxv. Antebellum planters had not yet undergone one key process of modernization, the devaluation of kinship ties and the replacement of kin groups with the nuclear family. See Andrejs Plakans and Charles Wetherell, "The Kinship Domain in an East European Peasant Community: Pinkenhof, 1833–1850," *American Historical Review* 93 (April 1988): 359–86.

Scholars now know that mobility was high in premodern societies and that theories of the mobility transition were exaggerated, but mobility nonetheless *felt* different in premodern societies; see Zelinsky, "Hypothesis of the Mobility Transition," 219–49. Historians and historical geographers are increasingly interested in the ways that women move through geographic space; see Linda K. Kerber, "Separate Spheres, Female Worlds, Woman's Place: The Rhetoric of Women's History," *Journal of American History* 75 (June 1988): 31–39; Gillian Rose and Miles Osborn, "Feminism and Historical Geography," *Journal of Historical Geography* 14 (October 1988): 405–9. For a creative discussion of how the sexes experience space in urban areas, see Dolores Hayden, *Redesigning the American Dream: The Future of Housing, Work, and Family Life* (New York: W. W. Norton & Company, 1984).

29. John J. Ambler, Jr., "Memoranda," 260, Ambler and Barbour Family Papers, UVA.

30. "J. E. B. Stuart's Letters to His Hairston Kin, 1850–1855," ed. Peter W. Hairston, *North Carolina Historical Review* 51 (July 1974): 288, J. E. B. Stuart to Bettie Hairston, 20 December 1852; Elisabeth Muhlenfeld, *Mary Boykin Chesnut: A Biography*, Southern Biography Series, ed. William J. Cooper, Jr. (Baton Rouge: Louisiana State University Press, 1981), 15; *Calhoun Papers*, ed. Hemphill, 8:525, William H. Fitzhugh to John C. Calhoun, 7 February 1824.

31. Reuben Grigsby and Reubenia Grigsby to Abner J. Grigsby, 25 May 1850, Grigsby Family Papers, VHS; Richard Beale Davis, "The Ball Papers: A Pattern of Life in the Low Country, 1800–1825," *South Carolina Historical Magazine* 65 (January 1964): 10–11; "Kitten" to Robert F. Hackett, 17 December 1853, Gordon

and Hackett Family Papers, UNC. On length of visits, see "Louisa" to Julia Pickens, 3 January 1833, Chiliab Smith Howe Papers, UNC; *Papers of John C. Calhoun*, Volume 1, *1801–1817*, ed. Robert L. Meriwether (Columbia: published by University of South Carolina Press for South Caroliniana Society, 1959), 51–52, John C. Calhoun to Floride Colhoun, 27 July 1810; *Papers of John C. Calhoun*, Volume 10, *1825–1829*, ed. Clyde N. Wilson and W. Edwin Hemphill (Columbia: published by University of South Carolina Press for South Carolina Department of Archives and History and the South Caroliniana Society, 1977), 550, John C. Calhoun to Patrick Noble, 10 January 1829; *Calhoun Papers*, ed. Wilson, 11:12, John C. Calhoun to James Edward Colhoun, 16 March 1829. Cf. Clinton, *Plantation Mistress*, 53–54, who argues that parents exchanged daughters but not sons.

32. Stephen D. Miller to "My Dear Daughter," 28 July 1835, Williams-Chesnut-Manning Papers, USC; Carrie and Addie Dogan to Caroline Gordon, 14 February 1848, Gordon and Hackett Family Papers, UNC. Grant McCracken, "The Exchange of Children in Tudor England: An Anthropological Phenomenon in Historical Context," *Journal of Family History* 8 (Winter 1983): 303–13, argues that an important phase of the socialization of Tudor children occurred while they were servants in other households.

33. Dabney Manuscript, 37, UVA; Hickerson, *Echoes of Happy Valley*, 10; *Calhoun Papers*, ed. Meriwether, 1:7–9, John C. Calhoun to Andrew Pickens, Jr., 21 January 1803; *Calhoun Papers*, ed. Hemphill, 8:75–76, John C. Calhoun to John Ewing Colhoun, 27 May 1823. See also Ernest McPherson Lander, Jr., *The Calhoun Family and Thomas Green Clemson: The Decline of a Southern Patriarchy* (Columbia: University of South Carolina Press, 1983), 73.

34. Peter G. Filene defines sex roles as the "behavior and attitudes . . . that are expected of an individual as a member of his or her gender." See *Him/Her/Self: Sex Roles in Modern America*, (Baltimore: Johns Hopkins University Press, 1986), xiii.

35. Dabney Manuscript, 35–36, UVA; Middleton, *Life in Carolina*, Part 3, "Reminiscences," by Nathaniel Russell Middleton, 186–88; DuBose, "Memoir," 1:145, John Witherspoon DuBose Papers, ALA; Hickerson, *Echoes of Happy Valley*, 36; Dabney Manuscript, 51, UVA.

36. John P. Belk to Oliver P. Hackett, 6 March 1832, Gordon and Hackett Family Papers, UNC.

37. It is impossible to identify these men in the census returns from 1810 and 1820 because the age categories—sixteen to twenty-six years, twenty-six to forty-five years—are so imprecise. These figures are only suggestive, indicating something of the nature of the dependency of planters' sons.

38. "Son" [probably William Dinwiddie] to William W. Dinwiddie, 4 March 1851, Dinwiddie Family Papers, UVA; Sarah Brown to Allen Brown, 22 February 1853, Hamilton Brown Papers, UNC. Joseph F. Kett, *Rites of Passage: Adolescence in America, 1790 to the Present* (New York: Basic Books, 1977), 14–31, uses the term "semi-dependency" to describe this phase in the lives of Northern men who lived between 1790 and 1840. As I read the evidence, young men did not come of age, in the sense of achieving autonomy, during their years at academies or colleges, as argued by Steven M. Stowe, *Intimacy and Power in the Old South: Ritual in the Lives of Planters*, New Studies in American Intellectual and Cultural History,

Thomas Bender, consulting editor (Baltimore: Johns Hopkins University Press, 1987), 122–59, and Jon L. Wakelyn, "Antebellum College Life and the Relations between Fathers and Sons," in *The Web of Southern Social Relations: Women, Family, and Education*, ed. Walter J. Fraser, Jr., R. Frank Saunders, Jr., and Jon L. Wakelyn (Athens: University of Georgia Press, 1985), 107–26. Nor do I see a "sharp discontinuity" between childhood and adulthood, as Dickson D. Bruce, Jr., *Violence and Culture in the Antebellum South* (Austin: University of Texas Press, 1979), 54, argues.

39. R. R. Hackett to R. F. Hackett, 5 February 1847, Gordon and Hackett Family Papers, UNC; *Calhoun Papers*, ed. Wilson, 11:84, John C. Calhoun to Armistead Burt, 6 November 1829, 11:259–60, John C. Calhoun to Armistead Burt, 11 November 1830; Addie and Caroline Dogan to Caroline Gordon, 28 December 1849, Gordon and Hackett Family Papers, UNC; Muhlenfeld, *Chesnut*, 41; Bertram Wyatt-Brown, *Southern Honor: Ethics and Behavior in the Old South* (New York: Oxford University Press, 1982), 5–6, 175–98.

40. John Cunningham to Benjamin C. Yancey, 1 January 1837, Benjamin Cudworth Yancey Papers, UNC; James W. Davis to John C. Davis, 11 December 1855, Conway, Black, and Davis Family Papers, USC; Louis T. Wigfall to L. Cheves, 8 August 1836, Langdon Cheves III Papers, SCHS. Regarding the eighteenth-century planter family, Philip Greven portrays fathers as strong-willed, remote individuals who had emotionally distant relationships with their children. See *The Protestant Temperament: Patterns of Child-Rearing, Religious Experience and the Self in Early America* (New York: Alfred A. Knopf, 1977), 274–81.

41. Jeremiah Choice to Franklin Townes, 14 May 1833, Townes Family Papers, USC; DuBose, "Memoir," 2:150–53, 177–89, John Witherspoon DuBose Papers, ALA.

42. On the upbringing of daughters, see Stowe, *Intimacy and Power*, 134, 137–50; Clinton, *Plantation Mistress*, 45–49, 123–38; Scott, *Southern Lady*, 7, 68–77. On property rights, see Shammas, Salmon, and Dahlin, *Inheritance in America*, 6, 83; Suzanne D. Lebsock, "Radical Reconstruction and the Property Rights of Southern Women," *Journal of Southern History* 42 (May 1977): 196–97, 202, 208.

43. Addie and Caroline Dogan to Caroline Gordon, 21 April 1854, Gordon and Hackett Family Papers, UNC. On skills, see George Lee Simpson, Jr., *The Cokers of Carolina: A Social Biography of a Family* (Chapel Hill: published for Institute for Research in Social Science by University of North Carolina Press, 1956), 37–38, 44; Muhlenfeld, *Chesnut*, 14–16.

44. *Calhoun Papers*, ed. Wilson, 16:519, Anna Calhoun Clemson to Patrick Calhoun, 29 October 1842, 560, Anna Calhoun Clemson to Patrick Calhoun, 3 December 1842; Lander, *Calhoun Family*, 55; A. L. Hackett to R. F. Hackett, 23 January 1847, Gordon and Hackett Family Papers, UNC.

45. Censer, *Planters and Children*, 89–91; Lewis, *Pursuit*, 188–208; Smith, *Great House*, 126–50; Narrative of Jane Harris Woodruff, 298–300, Harris Papers, SCHS. Other scholars stress the involvement of parents in their children's choices and the importance of both affection and property considerations to parents and children. See Stowe, *Intimacy and Power*, 96–106; Clinton, *Plantation Mistress*, 59–67; Wyatt-Brown, *Southern Honor*, 206–12; Wyatt-Brown, "Ideal Typology," 9; Scott, *South-*

ern Lady, 23–25; Guion Griffis Johnson, "Courtship and Marriage Customs in Ante-bellum North Carolina," *North Carolina Historical Review* 8 (October 1931): 384–86. Cf. Censer, *Planters and Children*, 65–95, who argues that children chose their own spouses with little input from their parents, who had socialized them so well that their choices usually pleased their parents. Stowe, *Intimacy and Power*, 157–59; Friedman, *Enclosed Garden*, 49–52; Censer, *Planters and Children*, 56–57, note that going away to school for the first time was an important transition in a young woman's life, as was a religious conversion.

46. Mary Ann Taylor Harwell to Marion Singleton, 3 May 1834, Singleton Family Papers, LC; Stowe, *Intimacy and Power*, 125–27; Clinton, *Plantation Mistress*, 37. Censer, *Planters and Children*, 88–90, argues that North Carolina planters tended to marry individuals who lived within a thirty-mile radius of home. Clinton, *Plantation Mistress*, 60, notes that planter women who married between 1765 and 1815 had a median age at first marriage of twenty years; the age for men was twenty-eight years. Censer, *Planters and Children*, 91–92, finds an average age and a median age at first marriage of twenty for selected North Carolina planters' daughters; for men, the average age was twenty-five and the median age was twenty-four. Burton, *In My Father's House*, 118, discovers an average age of twenty for women and twenty-five for men among whites of all social classes in Edgefield County, South Carolina, in 1860. Scott, *Southern Lady*, 25, believes that girls commonly married as teenagers.

47. Other historians argue that planter women were dependent on men. George C. Rable, *Civil Wars: Women and the Crisis of Southern Nationalism* (Urbana: University of Illinois Press, 1989), 1–30, and Fox-Genovese, *Within the Plantation Household*, 30, 101–2, 192–241, posit that women accepted or enjoyed their dependence, while Suzanne Lebsock, *The Free Women of Petersburg: Status and Culture in a Southern Town, 1784–1860* (New York: W. W. Norton & Company, 1984), xiii–xv, 15–53; Clinton, *Plantation Mistress*, 3–15, 164–79; and Scott, *Southern Lady*, 23–44, 46–79, tend to emphasize women's discontent with or struggles against dependence.

48. Eugene D. Genovese, *Roll, Jordan, Roll: The World the Slaves Made* (New York: Pantheon Books, 1974), 3–7, passim. In Part 1, Genovese includes examples from the entire South, while I am arguing that paternalistic behavior was more common among seaboard planters. James Oakes, *The Ruling Race: A History of American Slaveholders* (New York: Alfred A. Knopf, 1982), makes a somewhat similar argument, positing that paternalism disappeared from most of the planter class by the early nineteenth century, surviving longest in the oldest areas (the seaboard, the Gulf Coast, and the Mississippi Valley), to be replaced by impersonal, capitalistic race relations. (Oakes does not explore planter family life in depth, however.) Peter J. Parish, *Slavery: History and Historians* (New York: Harper & Row, 1989) discusses the debate on Genovese's work but omits most scholarship on slave and planter women. See also Jacqueline Jones, *Labor of Love, Labor of Sorrow: Black Women, Work, and the Family from Slavery to the Present* (New York: Basic Books, 1985), 11–43; Paula Giddings, *When and Where I Enter: The Impact of Black Women on Race and Sex in America* (New York: William Morrow and Company, 1984), 41–46; and *We Are Your Sisters: Black Women in the Nine-*

teenth Century, ed. Dorothy Sterling (New York: W. W. Norton & Company, 1984), 1–84, who discuss the exploitation of slave women and their resistance to it. See below for a discussion of planter women and slaves.

49. *The Ideology of Slavery: Proslavery Thought in the Antebellum South, 1830–1860*, ed. Drew Gilpin Faust (Baton Rouge: Louisiana State University Press, 1981), 64–68; George M. Fredrickson, *The Black Image in the White Mind: The Debate on Afro-American Character and Destiny, 1817–1914* (New York: Harper & Row, 1971), 55–68; Richard S. Dunn, "A Tale of Two Plantations: Slave Life at Mesopotamia in Jamaica and Mount Airy in Virginia, 1799 to 1828," *William and Mary Quarterly*, 3d ser., 34 (January 1977): 43; H. H. Townes to Rachel Townes, 28 October 1845, H. H. Townes to J. A. Townes, 14 March 1836, H. H. Townes to G. F. Townes, 1 March 1832, H. H. Townes to Rachel Townes, 20 November 1834, Townes Family Papers, USC. John C. Inscoe found similar instances of paternalism; see *Mountain Masters, Slavery, and the Sectional Crisis in Western North Carolina* (Knoxville: University of Tennessee Press, 1989), 93–98.

50. Fischer, *Albion's Seed*, 279, among other scholars, points out that the family was also a sphere of authority.

51. Genovese, *Roll, Jordan, Roll*, 81–84; Todd L. Savitt, *Medicine and Slavery: The Diseases and Health Care of Blacks in Antebellum Virginia* (Urbana: University of Illinois Press, 1978), 160–61; Muhlenfeld, *Chesnut*, 15; Rebecca Sims to George C. Dromgoole, 22 February 1844, Edward Dromgoole Papers, UNC; Mary Bunting to David J. Bunting, 18 February 1845, Mary Bunting Manuscript, USC; "Hattie" [Dogan?] to R. F. Hackett, 17 August 1854, Gordon and Hackett Family Papers, UNC. On managing plantations and the physical care of slaves, see Clinton, *Plantation Mistress*, 16–29, and Scott, *Southern Lady*, 28–33, 36–37.

52. Samuel Townes to Rachel Townes, 9 March 1829, Townes Family Papers, USC. Historians probably need a word other than *paternalism* that reflects the experiences of planter women and planter men, but this book will continue to use the term. Lebsock, *Free Women of Petersburg*, 136–41, shows that more white women than white men freed slaves in antebellum Petersburg, Virginia; she explains this as a manifestation of women's attachments to individual slaves, which was a "subversive influence" although it did not constitute a full-blown attack on slavery. Clinton, *Plantation Mistress*, 180–98, discusses women's "paradoxical role"; they were "traditional nurturers," but they also saw slaves as economic commodities and blamed slaves for their own heavy work duties. (Friedman, *Enclosed Garden*, 81–82, mentions cases of white women mistreating black women.) Scott, *Southern Lady*, 46–48, describes the plantation mistress as an "arbiter" of relationships and describes the bonds, fraught with tension and sometimes affection, between mistresses and house slaves.

Rable, *Civil Wars*, 31–37, argues that planter women shared planter men's racism and often mistreated slaves; he posits that planter women who were kind to slaves merely "strengthened slavery by concealing . . . its oppressive character beneath a veneer of compassion." (Surely slaves preferred kind treatment, regardless of planter women's motives.) Fox-Genovese raises many interesting questions about race, class, and gender in *Within the Plantation Household*, but it is difficult to accept her contention that planter women were "more crudely racist" (35) than planter men because she does not explore the racial attitudes of planter men.

53. James P. Cocke to Richard Archer, 25 April 1853, Archer Papers, UT; Hickerson, *Echoes of Happy Valley*, 64; "Adelaide" [Stokes] to R. F. Hackett, 7 January 1846, Gordon and Hackett Family Papers, UNC.

54. *"Journal of a Secesh Lady,"* ed. Crabtree and Patton, 5, 15 July 1860; Hunter Nicholson to R. F. Hackett, 17 May 1854, Gordon and Hackett Family Papers, UNC; Hickerson, *Happy Valley*, 59–63. On the experience and love of place, see David E. Sopher, "The Landscape of Home: Myth, Experience, Social Meaning," in *The Interpretation of Ordinary Landscapes: Geographical Essays*, ed. D. W. Meinig (New York: Oxford University Press, 1979), 137–38, and works by Yi-Fu Tuan, especially *Space and Place: The Perspective of Experience* (Minneapolis: University of Minnesota Press, 1977) and "Place: An Experiential Perspective," *Geographical Review* 65 (April 1975): 152. Love of place had been a basic component of Anglo-American and African-American culture since the colonial period; see Mechal Sobel, *The World They Made Together: Black and White Values in Eighteenth-Century Virginia* (Princeton, N.J.: Princeton University Press, 1987), 71, 74–75.

55. Kenzer, *Kinship and Neighborhood*, 29–51; Bruce Collins, *White Society in the Antebellum South*, Studies in Modern History, general eds. John Morrill and David Cannadine (London: Longman Group Limited, 1985), 138–41; Wyatt-Brown, *Southern Honor*, and "Ideal Typology," 4–5, 22; Wilbur J. Cash, *The Mind of the South* (New York: Alfred A. Knopf, 1941), all argue that kinship muted class distinctions, while Frank L. Owsley, *Plain Folk of the Old South*, with a new introduction by Grady McWhiney (Baton Rouge: Louisiana State University Press, 1949, 1982), argues for yeoman self-sufficiency. Harris, *Plain Folk and Gentry*, 5–6, argues that participation in the cotton economy and hatred for blacks overshadowed class differences among whites. Genovese, *Roll, Jordan, Roll* and *Political Economy of Slavery: Studies in the Economy and Society of the Slave South* (New York: Pantheon Books, 1965), contends that the slaveowning elite dominated all whites, enforcing a hegemony that was less obvious than but just as effective as its domination of slaves.

56. Dabney Manuscript, 34–35, UVA; Addie and Caroline Dogan to Caroline Gordon, 26 November 1852, 5 October 1853, 21 April 1854, Gordon and Hackett Family Papers, UNC.

57. Charles A. L. Lewis to Robert M. T. Hunter, 29 November 1832, 30 November 1832, R. M. T. Hunter Papers, UVA; John B. Edmunds, Jr., *Francis W. Pickens and the Politics of Destruction*, Fred W. Morrison Series in Southern Studies (Chapel Hill: University of North Carolina Press, 1986), 99, 104–6, 128–29; Burton, *In My Father's House*, 70, 150, 159–60; Fifth Decennial Census of the United States, South Carolina, Abbeville County, 100; Sixth Decennial Census of the United States, South Carolina, Pickens County, 354; Seventh Decennial Census of the United States, Slave Schedule, South Carolina, Pickens County, n.p. (entry for Floride Calhoun); Lander, *Calhoun Family*, 98–100, 103–4, 113.

58. John Cooke to "My Dear Niece," 9 January 1835, Noland Family Papers, UVA; Dabney Manuscript, 3, 2, UVA.

Chapter 2. In Search of Manly Independence: The Migration Decision

1. Other scholars have discussed the high mobility of the Southern white population and its commitment to change, but they have overlooked or downplayed these differences between generations of men and between the sexes. See Bruce Collins, *White Society in the Antebellum South*, Studies in Modern History, general eds. John Morrill and David Cannadine (London: Longman Group Limited, 1985), and James Oakes, *The Ruling Race: A History of American Slaveholders* (New York: Alfred A. Knopf, 1982), who captures especially well the acquisitiveness of many white male migrants (73–76). The rhetoric of "independence" appeared among men who came of age in other parts of Jacksonian America and may have been part of a national trend toward more modern, capitalistic, individualistic behavior, but the dynamics of the planter family and the plantation economy gave the idea a special urgency for these young men.

2. S. A. Townes to J. A. Townes, 1 October 1834, Townes Family Papers, USC. See Table 2. These ages are representative of other men who moved to the Southwest between 1830 and 1840. See Peter D. McClelland and Richard J. Zeckhauser, *Democratic Dimensions of the New Republic: American Interregional Migration, Vital Statistics, and Manumissions, 1800–1860* (Cambridge: Cambridge University Press, 1982), 141. Men who migrated to other American frontiers were usually in their twenties or early thirties; see Lillian Schlissel, *Women's Diaries of the Westward Journey* (New York: Schocken Books, 1982), 28; John Mack Faragher, *Women and Men on the Overland Trail* (New Haven: Yale University Press, 1979), 17.

3. See Table 2. William C. Carrington to Louisa C. Carrington, 20 April 1841, Carrington Family Papers, VHS. John J. Ambler, Jr., "Memoranda," 232, Ambler and Barbour Family Papers, UVA. See also H. L. Benning to Benjamin C. Yancey, 3 December 1832, Benjamin Cudworth Yancey Papers, UNC; Edmund Scarborough to Samuel Scarborough, 13 December 1840, Scarborough Family Papers, DU.

4. John Hebron Moore, *The Emergence of the Cotton Kingdom in the Old Southwest: Mississippi, 1770–1860* (Baton Rouge: Louisiana State University Press, 1988), 6; Gavin Wright, *The Political Economy of the Cotton South: Households, Markets, and Wealth in the Nineteenth Century* (New York: W. W. Norton & Company, 1978), 103; Robert William Fogel and Stanley L. Engerman, *Time on the Cross: The Economics of American Negro Slavery*, 2 vols. (Boston: Little, Brown and Company, 1974), 1:196–209; Alfred Glaze Smith, Jr., *Economic Readjustment of an Old Cotton State: South Carolina, 1820–1860* (Columbia: University of South Carolina Press, 1958), 91, 95; Lewis Cecil Gray, *History of the Agriculture of the Southern United States to 1860*, 2 vols. (Baltimore: Waverly Press, 1933), 2:912; Avery Odell Craven, *Soil Exhaustion as a Factor in the Agricultural History of Virginia and Maryland, 1601–1860* (Urbana: University of Illinois Press, 1925), 9, 118–20, 122–26. Wright, *Political Economy*, 17, argues that Southerners exaggerated the extent of soil exhaustion, and idem, *Old South, New South: Revolutions in the Southern Economy Since the Civil War* (New York: Basic Books, 1986), 35–36, notes that Southerners used fertilizers to raise cotton in the allegedly exhausted seaboard after the Civil War. Men may have believed that the large scale of plantations in the Southwest made them more profitable. See Chapter 3.

5. John J. Ambler, Jr., "Memoranda," 308, Ambler and Barbour Family Papers, UVA. See also H. L. Benning to Benjamin C. Yancey, 3 December 1832, Benjamin Cudworth Yancey Papers, UNC; Quintus Barbour to Elizabeth B. Ambler, 12 February 1835, Ambler and Barbour Family Papers, UVA; Louis T. Wigfall to Langdon Cheves II, 8 August 1836, Langdon Cheves III Papers, SCHS; A. L. Hackett to R. F. Hackett, 23 January 1847, N. N. Fleming to R. F. Hackett, 11 January 1851, J. L. Moseley to R. F. Hackett, 22 February 1851, Gordon and Hackett Family Papers, UNC; Henry Tayloe to Benjamin Tayloe, 11 August 1837, Tayloe Family Papers, UVA. Planters' sons rarely mentioned political honors as a goal.

6. Lee Soltow, *Men and Wealth in the United States, 1850–1870* (New Haven: Yale University Press, 1975), 176, 180; Thomas Felix Hickerson, *Happy Valley: History and Genealogy* (Chapel Hill: by the author, 1940), 116; Joseph H. Dukes to John L. Manning, 10 July 1839, Chesnut-Miller-Manning Papers, USC.

7. Elliott J. Gorn, *The Manly Art: Bare-Knuckle Prize Fighting in America* (Ithaca: Cornell University Press, 1986), 140–41; J. William Harris, *Plain Folk and Gentry in a Slave Society: White Liberty and Black Slavery in Augusta's Hinterlands* (Middletown, Conn.: Wesleyan University Press, 1985), 21–22; Edward L. Ayers, *Vengeance and Justice: Crime and Punishment in the Nineteenth-Century American South* (New York: Oxford University Press, 1984), 41–42; Rhys Isaac, *The Transformation of Virginia, 1740–1790* (Chapel Hill: published for Institute of Early American History and Culture by University of North Carolina Press, 1982), 131–32.

Historians are still exploring the impact of this view of independence on family life. Jan Lewis, *The Pursuit of Happiness: Family and Values in Jefferson's Virginia* (Cambridge: Cambridge University Press, 1983), 1–9, 106–68, sees its roots in pre-Revolutionary republican ideas but notes that by the early nineteenth century many young men expected parents to support them and give them "independence." Drew Gilpin Faust, *James Henry Hammond and the Old South: A Design for Mastery* (Baton Rouge: Louisiana State University Press, 1982), 41–43, notes that inevitable tensions existed between the idea of republican independence and the realities of South Carolina's social and economic system. Wright, *Political Economy*, 46, stresses the purely economic benefits of owning a farm of one's own. Historians have applied the concept of "family strategies" to such family decisions, emphasizing purely economic motives; see, for example, Claudia Goldin, "Family Strategies and the Family Economy in the Late Nineteenth Century: The Role of Secondary Workers," in *Philadelphia: Work, Space, Family, and Group Experience in the Nineteenth Century*, ed. Theodore Hershberg (New York: Oxford University Press, 1981), 277–310. The motives of planters' sons were too complicated, however, to be called purely economic.

8. James Berryman to Newton Berryman, 5 December 1834, Berryman Family Papers, VHS; Willliam M. Ambler to John J. Ambler, Jr., 23 August 1834, Ambler and Barbour Family Papers, UVA; Louis T. Wigfall to John L. Manning, 18 January n.d. [1839], Williams-Chesnut-Manning Papers, USC. On independence and slavery, see Isaac, *Transformation*, 136; James L. Roark, *Masters Without Slaves: Southern Planters in the Civil War and Reconstruction* (New York: W. W.

Norton & Company, 1977), 68–108; Eugene D. Genovese, *Political Economy of Slavery: Studies in the Economy and Society of the Slave South* (New York: Pantheon Books, 1965), 31–34. Men in their twenties and thirties usually articulated these ideas, but not every young man agreed with them, and some middle-aged and older men also embraced these ideas. Kenneth S. Greenberg, *Masters and Statesmen: The Political Culture of American Slavery*, New Studies in American Intellectual and Cultural History, Thomas Bender, consulting editor (Baltimore: Johns Hopkins University Press, 1985), 86–87, notes that Southern politicians throughout the antebellum era expressed a fear of enslavement by political foes.

9. "A Journey through the South in 1836: Diary of James D. Davidson," ed. Herbert A. Kellar, *Journal of Southern History* 1 (February–November 1935): 360, 12 November 1836; Orville Vernon Burton, *In My Father's House Are Many Mansions: Family and Community in Edgefield, South Carolina*, Fred W. Morrison Series in Southern Studies (Chapel Hill: University of North Carolina Press, 1985), 65, 69, 144; Cloud Memoirs, 3, Isaac Newton Cloud Memoirs, UVA; Kenneth McKenzie to Duncan McLaurin, n.d. October 1848, Duncan McLaurin Papers, DU.

10. Louis T. Wigfall to John L. Manning, 18 January n.d. [1830s], 22 February n.d. [1830s], Williams-Chesnut-Manning Papers, USC; Joshua Reynolds to John B. Miller, 7 August 1826, Miller-Furman-Dabbs Papers, USC. See also Thomas Brown, "An Account of the Lineage of the Brown Family," 85–88, Ambler-Brown Family Papers, DU. Louis Wigfall's parents died by the time he was a teenager and left him a sizable inheritance, but he nonetheless felt oppressed by obligations to his older brother Arthur and a "blackguard" cousin who was trying to "ruin" him. See Louis T. Wigfall to John L. Manning, 22 February n.d. [1830s], Williams-Chesnut-Manning Papers, USC.

11. S. A. Townes to G. F. Townes, 3 January 1832, Townes Family Papers, USC; William C. Carrington to Louisa E. Carrington, 25 December 1840, Carrington Family Papers, VHS; Henry E. Blair to Jane I. Blair, 9 October 1849, Blair Papers, VHS.

12. Henry E. Blair to Jane I. Blair, 9 October 1849, Blair Papers, VHS; H. H. Townes to G. F. Townes, 16 January 1834, Townes Family Papers, USC; William C. Carrington to Louisa E. Carrington, 20 April 1841, Carrington Family Papers, VHS; W. B. Blake to M. P. Blue, 8 July 1858, Matthew P. Blue Papers, ALA. Annette Kolodny, *The Land Before Her: Fantasy and Experience of the American Frontiers* (Chapel Hill: University of North Carolina Press, 1984), 3, discusses the sexual imagery used by European and American men to describe the New World and the American West. William Gilmore Simms noted the contrast between the coarse, masculine Southwest, where there were no restraints on human passions, and the genteel, feminine seaboard, a theme that appeared in his personal life and his fiction. See John McCardell, "Poetry and the Practical: William Gilmore Simms," in *Intellectual Life in Antebellum Charleston*, ed. Michael O'Brien and David Moltke-Hansen (Knoxville: University of Tennessee Press, 1986), 188–90.

13. Dabney Manuscript, 146, UVA; John W. DuBose, "Memoir of Four Families," 2:128–29. John Witherspoon DuBose Papers, ALA; Louisa Cunningham to Benjamin Yancey, 6 August 1833, Benjamin Cudworth Yancey Papers, UNC;

William Whitfield to Gaius Whitfield, 2 March 1837, Gaius Whitfield Papers, ALA.

14. Ellen Hard Townes, "Genealogy of the Townes Family," 1–2; H. H. Townes to G. F. Townes, 4 August 1831, H. H. Townes to J. A. Townes, 19 May 1833, H. H. Townes to Rachel Townes, 8 December 1833, H. H. Townes to W. A. Townes, 21 April 1833, H. H. Townes to G. F. Townes, 27 August n.d. [1833], 19 December 1840, Lucretia and H. H. Townes to Rachel Townes, 8 November 1839, H. H. Townes to W. A. Townes, 18 August 1841, H. H. Townes to Rachel Townes, 30 November 1843, H. H. Townes to John Townes, 2 January 1845, Townes Family Papers, USC; Fifth Decennial Census of the United States, South Carolina, Greenville County, 336; Sixth Decennial Census of the United States, South Carolina, Abbeville County, 6. Steven M. Stowe, *Intimacy and Power in the Old South: Ritual in the Lives of Planters* (Baltimore: Johns Hopkins University Press, 1987), 28–30, describes the dispute with Benjamin Perry but identifies Samuel as the oldest son.

Henry and Samuel Townes fit the profiles of conservative firstborn sons and innovative laterborn sons described by psychologists; see Stephen P. Bank and Michael D. Kahn, *The Sibling Bond* (New York: Basic Books, 1982); Alfred Adler, *Understanding Human Nature*, trans. W. Beran Wolfe (Garden City, N.Y.: Garden City Publishing, 1927). Birth order must have affected the choices of many men, since nine of the twenty firstborn sons (45 percent) in Table 2 migrated, while forty of the sixty-six laterborn sons (61 percent) migrated, but birth order alone cannot explain these decisions.

15. H. L. Benning to Benjamin C. Yancey, 3 December 1832, Benjamin Cudworth Yancey Papers, UNC; J. W. Carrigan to William A. Carrigan, 23 September 1851, Carrigan Papers, USC; Journal of William A. Lenoir, 3 November 1836, vol. 155, Lenoir Family Papers, UNC; Richard C. Ambler to John J. Ambler, Jr., 2 February 1835, Ambler and Barbour Family Papers, UVA; Davis R. Dewey, *State Banking Before the Civil War* (Washington, D.C.: Government Printing Office, 1910), 184, 189. Bertram Wyatt-Brown, *Southern Honor: Ethics and Behavior in the Old South* (New York: Oxford University Press, 1982), 183–84, points out that many banks were controlled by a few powerful families.

16. See Table 2. I describe slaveholding rather than landholding because wills or inventories usually provide numbers of slaves but not always the acreage or value of land bequests. Burton, *In My Father's House*, 108, suggests that inequitable wills resulted in smaller bequests of land (not slaves) for planter sons, which motivated them to go west. William Bullock's will is illegible; John Herbert Dent and Abraham Nott left no surviving wills or probate records. John C. Calhoun also died intestate, but his son Andrew bought out most of the other heirs immediately leaving them one or two slaves each at most. See Court Order regarding estate of Peter F. Archer, 17 August 1814, Powhatan County Order Book No. 12, 200, Powhatan County Courthouse; Will of William Bullock, November 1829, Granville County Will Book II, 253–57, NCA; Ernest McPherson Lander, Jr., *The Calhoun Family and Thomas Green Clemson: The Decline of a Southern Patriarchy* (Columbia: University of South Carolina Press, 1983), 132–137; Will of Micajah Carr, Albemarle County Will Book 5, 203, VSA; Will of James Hackett, November 1845,

Superior Court of Wilkes County, North Carolina, microfilm reel 289, 311, NCA; Will of William Polk, written 24 August 1833, probated February 1834, Wake County Will Book Y, 46–51, NCA; Will of John Tayloe III, 1824, 1825, 1827, Tayloe Family Papers, UVA; Will of Samuel Townes, 3 September 1825, Townes Family Papers, USC; Will of Needham Whitfield, written 20 March 1812, Wayne County Wills, 1776–1923, North Carolina, NCA. John Tayloe's will bequeaths slaves by plantation but does not give the exact number of slaves; Henry's share is an estimate based on various divisions in the will. Archer's sons each inherited two slaves; Carr's sons, one slave; Hackett's sons, two slaves; Townes's sons, from two to five slaves; and Whitfield's sons, seven slaves.

The legal codes of Virginia, North Carolina, and South Carolina barred primogeniture and entail for intestates by 1791; see *The Code of Virginia*, 2 vols. (Richmond: William F. Ritchie, 1849), 2:522n; *Laws of the State of North Carolina*, rev., by Hen. Potter, J. L. Taylor, and Bart. Yancey, 2 vols. (Raleigh: J. Gales, 1821), 1:465, 467; *The Statutes at Large of South Carolina*, ed. Thomas Cooper, 10 vols. (Columbia: A. S. Johnston, 1837), 5:572. Most scholars of inheritance practices stress the equal distribution of property among male children; see Burton, *In My Father's House*, 107–8; Jane Turner Censer, *North Carolina Planters and Their Children* (Baton Rouge: Louisiana State University Press, 1984), 108–18; Suzanne Lebsock, *The Free Women of Petersburg: Status and Culture in a Southern Town, 1784–1860* (New York: W. W. Norton & Company, 1984), 115; Carole Shammas, Marylynn Salmon, and Michel Dahlin, *Inheritance in America from Colonial Times to the Present* (New Brunswick: Rutgers University Press, 1987), 63.

17. See Table 2. William H. McNeill, "Introduction," in *Human Migration: Patterns and Policies*, ed. William H. McNeill and Ruth S. Adams (Bloomington: Indiana University Press, 1978), xi; Martin E. Marty, "Migration: The Moral Framework," in *Human Migration*, 397. Henry A. Tayloe apparently bought land in Alabama jointly with three of his brothers; see Indenture between Henry A. Tayloe, Benjamin O. Tayloe, and William H. Tayloe, 21 July 1834, Tayloe Family Papers, UVA.

18. Robert V. Wells, *Revolutions in Americans' Lives: A Demographic Perspective on the History of Americans, Their Families and Their Society*, Contributions in Family Studies, No. 6 (Westport, Conn.: Greenwood Press, 1982), 29–30, 55–56; Robert W. Fogel et al., "The Economics of Mortality in North America, 1690–1910: A Description of a Research Project," *Historical Methods: A Journal of Quantitative and Interdisciplinary History* 11 (1978):81; McClelland and Zeckhauser, *Demographic Dimensions*, 6–7, 51. On the different experiences of generations, see Alan B. Spitzer, "The Historical Problem of Generations," *American Historical Review* 78 (December 1973): 1353–85. In his fascinating study of agricultural reformer Edmund Ruffin, David F. Allmendinger, Jr., suggests that Ruffin developed new farming methods in Virginia partly because he had no older male relatives to instruct him in traditional methods; see *Ruffin: Family and Reform in the Old South* (New York: Oxford University Press, 1990), 26–27.

This profile is based on biographical information in Table 2 and the manuscript collections in this study. The twenty fathers in Table 2 owned a total of 1,896 slaves, based upon their peak slaveholdings, an average of 95 slaves per slaveowner. If

John Tayloe, who owned as many as 780 slaves, is omitted, the remaining nineteen fathers owned an average of 59 slaves, which is closer to the average of 53 slaves for the seaboard men in Table 1. The large number of children in the families in Table 2 (an average of eight) no doubt worsened the problem of providing for children; this was higher than the national average of six children for women who married in the late eighteenth century. See Wells, *Revolutions*, 92.

19. Gray, *Agriculture*, 1:541–42, 2:666; William Kauffman Scarborough, *The Overseer: Plantation Management in the Old South* (Baton Rouge: Louisiana State University Press, 1966), 30. No figures exist on the costs of maintaining a frontier plantation before it was a profitable concern, but they must have been considerable. Censer, *Planters and Children*, 130–31, finds that one-third of her North Carolina planter families containing a migrant owned land in the Southwest, but she does not specify whether the migrants went directly to those lands or to other areas of the Southwest. Among the fathers in Table 2, only William Polk sent a son (Lucius) to manage a Southwestern plantation.

20. Fogel et al., "Economics of Mortality," 92; William Henry Brodnax to Robert Walker Withers, 24 August 1832, William Henry Brodnax Papers, VHS. Brodnax's brother died, however, before he could fulfill these expectations. Fogel and Engerman, *Time on the Cross*, 1:153–55, estimate that the "investment" a master made in a slave began to pay at the same age, twenty-seven. Before the Panic of 1837, good land in Charlotte County, Virginia, sold for ten to fifteen dollars an acre, while prime land in Alabama and Mississippi went for thirty-five to forty-five dollars an acre; see Gray, *Agriculture*, 2:644–45, 895.

21. Joseph H. Pleck and Elizabeth H. Pleck, *The American Man* (Englewood Cliffs, N.J.: Prentice Hall, 1980), 6, 10, argue that the eighteenth century's "agrarian patriarchy" faded in the North by 1820 but imply that it lasted longer in the South. They believe, however, that the authority of fathers eroded after the Revolution and cite the migration of sons as an example of this shift. Isaac, *Transformation*, 311–12, argues that the westward movement undermined the sense of community in late eighteenth-century Virginia. James A. Henretta, "Family and Farms: Mentalité in Pre-Industrial America," *William and Mary Quarterly* 35 (1978): 31, notes that migration posed a subtle challenge to a father's authority. Cf. Censer, *Planters and Children*, 118, who believes that planter parents accepted their adult children's desire for independence.

22. Other scholars describe this divergence in male sex roles, although in somewhat different terms. See Lewis, *Pursuit*, 9–10; Wells, *Revolutions*, 8–9; Robert J. Brugger, *Beverly Tucker: Heart over Head in the Old South* (Baltimore: Johns Hopkins University Press, 1978), 42; Wilbur J. Cash, *The Mind of the South* (New York: Alfred A. Knopf, 1941), 52–53. Cf. Stowe, *Intimacy and Power*, who skillfully describes the conservative, patriarchal masculinity prevalent among seaboard men but implies that it, like most aspects of planter culture, remained essentially unchanged throughout the antebellum era.

23. Allen Brown to Hamilton Brown, 28 August 1854, Hamilton Brown Papers, UNC; Bleser, ed., *Hammonds of Redcliffe*, 59.

24. Kenneth McKenzie to Duncan McLaurin and John M. Stewart, 8 November 1833, Duncan McKenzie to Duncan McLaurin, 5 July 1846, Duncan McLaurin

Papers, DU; Will of Bushrod Powell, n.d. October 1839, Fairfax County Will Book, Book T-1, 115–16, VSA.

25. William M. Ambler to John J. Ambler, Jr., 1 July 1835, Ambler and Barbour Family Papers, UVA.

26. T. P. Huger to Langdon Cheves I, 14 December 1846, Langdon Cheves I Papers, SCHS; Archie Vernon Huff, Jr., *Langdon Cheves of South Carolina* (Columbia: University of South Carolina Press, 1977), 133–34, 208–9, 211, 237, 239.

27. David Leech to Joseph A. Leech, 25 January 1819, John M. D. Lowry to Joseph A. Leech, 12 March 1829, Leech Papers, DU. See also Adam S. Dandridge to R. M. T. Hunter, 11 September 1859, R. M. T. Hunter Papers, UVA.

28. Will of William Polk, written 24 August 1833, probated February 1834, Wake County Will Book Y, 46–51, NCA; Joseph H. Parks, *General Leonidas Polk C.S.A., the Fighting Bishop* (Baton Rouge: Louisiana State University Press, 1962), 26, 42 n.54; John J. Ambler, Jr., "Memoranda," 17, Phillipa Barbour to Elizabeth B. Ambler, 17 August 1834, Katherine Ambler to "My Dear Sons," 5 May 1835, Ambler and Barbour Family Papers, UVA; Willis Lea to William Lea, 4 February 1826, 15 November 1841, 20 May 1844, Addison Lea to William Lea, 21 May 1836, Lea Family Papers, UNC; Lander, *Calhoun Family*, 7–12, 22–26, 36–37; *Papers of John C. Calhoun*, Volume 15, *1839–1841*, ed. Clyde N. Wilson (Columbia: published by University of South Carolina Press for South Carolina Department of Archives and History and South Caroliniana Society, 1983), 114–16, John C. Calhoun to Thomas G. Clemson, 22 February 1840.

29. *Papers of John C. Calhoun*, Volume 5, *1820–1821*, ed. W. Edwin Hemphill (Columbia: published by University of South Carolina Press for South Carolina Department of Archives and History and South Caroliniana Society, 1971), 131–33, John C. Calhoun to Charles Tait, 20 May 1820. Cf. John Niven, *John C. Calhoun and the Price of Union*, Southern Biography Series, ed. William J. Cooper, Jr. (Baton Rouge: Louisiana State University Press, 1988), who portrays Calhoun as an insecure, defensive man. Among the fathers in Table 2, only John C. Calhoun, William Lea, John J. Ambler, Sr., and William Polk assisted their sons in migrating.

30. Patience Laye to Mr. and Mrs. J. Oates, 18 November 1858, Oates Papers, USC. See also Martha Pickens to Ann Lenoir, 28 October 1814, Chiliab Smith Howe Papers, UNC; Sarah H. Gayle Diary, 20 June 1830, Bayne and Gayle Family Papers, UNC. Kolodny, *Land Before Her*, 93–94, notes that American women from all regions objected to migration because it separated them from kin and friends.

31. Catherine R. Patterson to Neill Kelly, 7 February 1840, John N. Kelly Papers, DU (first and second examples); Caroline Gordon to Allen Brown, n.d. 1854, Gordon and Hackett Family Papers, UNC. See Table 2.

32. Hickerson, *Happy Valley*, 113; *Memoirs of Mary A. Maverick, arranged by Mary A. Maverick and her son George Madison Maverick*, ed. Rena Maverick Green (San Antonio: Alamo Printing Company, 1921), 12; Louisa Cunningham to Benjamin Yancey, 18 September 1839, Benjamin Cudworth Yancey Papers, UNC. Kolodny, *Land Before Her*, 94–95, and Julie Roy Jeffrey, *Frontier Women: The Trans-Mississippi West, 1840–1880*, American Century Series (New York: Hill and Wang, 1979), 37, observe that women from the Northeast and Midwest compared migration to death.

33. Mrs. W. O. Absher, *Land Entry Book, Wilkes County, North Carolina, 1778–1781* (North Wilkesboro: Genealogical Society of Wilkes County, n.d.), land deed 28; Hickerson, *Happy Valley*, 67, 194; Sarah Brown to her son, 22 February 1853, Hamilton Brown Papers, UNC; S. A. [for Sarah Ann, called "Ann"] Finley to Caroline Gordon, 30 December 1847, James Gordon Hackett Papers, DU.

34. Jane Dunn to William E. Jones, 19 December 1852, William Edmondson Jones Papers, VHS; Olive Packard to Chilion Packard, 10 December 1823, Packard Family Papers, SCHS; H. H. Townes to Rachel Townes, 8 December 1833, Townes Family Papers, USC; Elisabeth Showalter Muhlenfeld, "Mary Boykin Chesnut: The Writer and Her Work" (Ph.D. dissertation, University of South Carolina, 1978), Appendix C, 824. Women's apprehensions about the Southwestern climate resemble the fears of many Britons about the unhealthy climate of the Southern colonies and the West Indies. See Karen Ordahl Kupperman, "Fear of Hot Climates in the Anglo-American Colonial Experience," *William and Mary Quarterly*, 3rd ser., 41 (April 1984): 213–40.

35. Ray Mathis, *John Horry Dent: South Carolina Aristocrat on the Alabama Frontier* (University: published under sponsorship of Historic Chattahoochee Commission of University of Alabama Press, 1979), 149; Frances C. Berkeley to Callender Noland, 24 September 1838, Noland Family Papers, UVA; Lydia Riddick to Charles Riddick, 11 November 1852, Charles C. Riddick Papers, UNC. Marty, "Migration: The Moral Framework," 387–403, discusses migration as a moral issue but does not distinguish between men's and women's views. Other American women were disinclined to believe claims about the riches to be had in the Southwest; see Kolodny, *Land Before Her*, 105. George C. Rable, *Civil Wars: Women and the Crisis of Southern Nationalism* (Urbana: University of Illinois Press, 1989), 29–30, remarks that many planter women adhered to "anticommercial values" in the late antebellum era, although he argues that they nonetheless supported the slavery regime and the Southern social system.

36. Willis Lea to William Lea, 11 January 1847, Lea Family Papers, UNC; *The Lides Go South . . . and West: the Records of a Planter Migration in 1835*, ed. Fletcher M. Green (Columbia: University of South Carolina Press, 1952), 15; Israel Pickens to Martha L. Pickens, 18 March 1816, Israel Pickens Papers, ALA. There is no evidence that the moral authority of the Southern "lady" enhanced women's actual influence in families considering migration. Nor did women's "separate sphere" or their allegedly more spiritual nature give them an effective voice in family discussions of migration; neither did the large amount of work they performed in raising children and running households. See William R. Taylor, *Cavalier and Yankee: The Old South and American National Character* (New York: Harper & Row, 1957; Harper Torchbook, 1969), 146–76; Catherine Clinton, *The Plantation Mistress: Woman's World in the Old South* (New York: Pantheon Books, 1982), 16–35; Anne Firor Scott, *The Southern Lady: From Pedestal to Politics, 1830–1930* (Chicago: University of Chicago Press, 1970), 4–21, 27–37. Cf. Lewis, *Pursuit*, 211, and Carl N. Degler, *At Odds: Women and the Family in America from the Revolution to the Present* (New York: Oxford University Press, 1980), 30–42, 50, who argue that separate spheres gave women genuine moral authority in the family. Lebsock, *Free Women of Petersburg*, xv–xvi, 232–36, argues

that the cult of womanhood and its "separate spheres" constituted a backlash against women's increasing autonomy.

Most scholars of Southern families agree that women were excluded from the decision to migrate; see Lewis, *Pursuit*, 144–45; Clinton, *Plantation Mistress*, 166–67; and Oakes, *Ruling Race*, 87–88. Cf. Censer, *Planters and Children*, 132, who stresses women's influence over their husbands. Historians of migration to other American frontiers disagree on women's exclusion from decision-making. Kolodny, *Land Before Her*, 31–34; Schlissel, *Women's Diaries*, 28, 31, 14; Faragher, *Overland Trail*, 67, 163, argue that women were excluded, while Sandra Myres, *Westering Women and the Frontier Experience, 1800–1915* (Albuquerque: University of New Mexico Press, 1982), 9–11, 16–36; Joanna L. Stratton, *Pioneer Women: Voices from the Kansas Frontier* (New York: Simon and Schuster, 1981), 44–45; Glenda Riley, *Frontierswomen: The Iowa Experience* (Ames: Iowa State University Press, 1981), 5–11, 26; Jeffrey, *Frontier Women*, 30–33, argue that women were involved in decision-making or approved the decision once it was made.

37. Gayle Diary, 26 October 1827, Bayne and Gayle Family Papers, UNC; Elizabeth B. Ambler to John J. Ambler, Jr., 25 September 1835, Ambler and Barbour Family Papers, UVA; Sarah Brown to Allen Brown, 22 February 1853, Hamilton Brown Papers, UNC. On the inequities of planter marriages, see Jean E. Friedman, *The Enclosed Garden: Women and Community in the Evangelical South, 1830–1900* (Chapel Hill: University of North Carolina Press, 1985), 21–38; Wyatt-Brown, *Southern Honor*, 254–71; Clinton, *Plantation Mistress*, 68–85; Eugene D. Genovese, *Roll, Jordan, Roll: The World the Slaves Made* (New York: Pantheon Books, 1974), 74; Anne Firor Scott, "Women's Perspective on the Patriarchy in the 1850s," *Journal of American History* 61 (June 1974): 55–64; Scott, *Southern Lady*, 46–66. Cf. Censer, *Planters and Children*, 65–95, who emphasizes parity and companionship in planter marriages.

38. Elizabeth M. Otey to James Otey, 23 February 1823, 21 September 1825, James Hervey Otey Papers, UNC; Sarah W. Irby to John W. Irby, 18 April 1849, Neblett and Irby Family Papers, UVA; Mary E. P. Allen to John J. Allen, Jr., 28 December 1854, 6 March 1855, Allen Family Papers, UVA.

39. *The American Slave: A Composite Autobiography*, ed. George P. Rawick, Contributions in Afro-American and African Studies, No. 11 (Westport, Conn.: Greenwood Publishing Company, 1972), *South Carolina*, 3: part 3, 74; William Townes to S. A. Townes, 1 May 1824, Townes Family Papers, USC; Samuel Van Wyck to Margaret Van Wyck, 28 March 1860, Maverick and Van Wyck Family Papers, USC. See also Parks, *Leonidas Polk*, 63–64.

40. S. A. Townes to G. F. Townes, 18 July 1833, 22 June 1834, H. H. Townes to G. F. Townes, 16 January 1834, Townes Family Papers, USC; DuBose, "Memoir," 2:191, John Witherspoon DuBose Papers, ALA.

41. Gayle Diary, 11 May 1830, 9 May 1830 or 9 June 1830, Bayne and Gayle Family Papers, UNC; Elizabeth Blaetterman to "Mrs. Michie," 26 December 1845, 8 July 1846, n.d., George Carr Papers, UVA. The dates and pagination in the Gayle Diary are sometimes inconsistent.

42. Mary E. P. Allen to John J. Allen, Jr., 6 March 1855, Allen Family Papers,

UVA; S. A. Townes to G. F. Townes, 18 July 1833, G. F. Townes to J. A. Townes, 4 February 1834, Townes Family Papers, USC; Bernard Carr to Mead Carr, 8 April 1829, George Carr Papers, UVA. Widows could have given property to their sons to help them migrate, depending upon restrictions on property disposition in their husbands' wills, but I have found no record of these kinds of transactions.

43. Allan Kulikoff, "Migration and Cultural Diffusion in Early America, 1600–1860: A Review Essay," *Historical Methods* 19 (Fall 1986): 164; John B. Boles, *Black Southerners, 1619–1869* (Lexington: University of Kentucky Press, 1983), 68; McClelland and Zeckhauser, *Demographic Dimensions*, 6, 159–64; Daniel M. Johnson and Rex R. Campbell, *Black Migration in America: A Social and Demographic History* (Durham: Duke University Press, 1981), 25.

44. Lawrence W. Levine, *Black Culture and Black Consciousness: Afro-American Folk Thought from Slavery to Freedom* (New York: Oxford University Press, published with aid of a Bicentennial Grant from Phi Beta Kappa Society, 1977), 14–15; Rawick, ed., *American Slave, South Carolina*, 3: part 3, 219; *Texas*, 5: part 3, 111; Mary A. Livermore, *The Story of My Life* (Hartford, Conn.: A. D. Worthington & Co., 1899), 232; Joseph Holt Ingraham, *The South-West. By a Yankee*, March of America Facsimile Series, No. 76 (New York: Harper & Brothers, 1835; repr. Ann Arbor, University Microfilms, 1966), 2:235; Herbert G. Gutman, *The Black Family in Slavery and Freedom, 1750–1925* (New York: Pantheon Books, 1976), 357–58; *We Are Your Sisters: Black Women in the Nineteenth Century*, ed. Dorothy Sterling (New York: W. W. Norton & Company, 1984), 43. Cf. Paul D. Escott, *Slavery Remembered: A Record of Twentieth-Century Slave Narratives* (Chapel Hill: University of North Carolina Press, 1979), 57–58, who argues that the narratives do not prove that slavery was more brutal in the Southwest, although he notes that some ex-slaves believed this to be the case.

45. Gutman, *Black Family*, 38, 43, finds similar variations in sex ratios for counties in Virginia and North Carolina in 1860, although none of these counties had ratios lower than 91.1 males for every 100 females. McClelland and Zeckhauser, *Demographic Dimensions*, 8, 52, 162, argue that slave migrants were equally divided by sex, while Johnson and Campbell, *Black Migration*, 27–28, maintain that females predominated in the slave force which planters brought to the Southwest. Boles, *Black Southerners*, 225, ably summarizes the debate on whether seaboard slaves were bred for the purposes of sale and concludes (68–70) that they were not. Table 3 contains some puzzles, notably the ratio of 125.2 in Sumter in 1820, the highest among all of the counties examined.

46. Livermore, *Story of My Life*, 327; Rawick, ed., *American Slave, Texas*, 4: part 1, 62–63; *Texas*, 5: part 3, 45–47; Boles, *Black Southerners*, 68–69. On the slave family, see Deborah Gray White, *Ar'n't I a Woman? Female Slaves in the Plantation South* (New York: W. W. Norton & Company, 1985); Gutman, *Black Family*; Genovese, *Roll, Jordan, Roll*. In 1829 Louisiana forbad the importation of slave children under the age of ten unless accompanied by their mothers "if living"; the law also prohibited the separation of children and mothers by sale within the state. It probably did little, however, to protect the slave family because it made no reference to fathers, and slave traders evaded the law by stating that children were

orphans. See Wendell Holmes Stephenson, *Isaac Franklin: Slave Trader and Planter of the Old South, with Plantation Records* (Baton Rouge: Louisiana State University Press, 1938), 72, 78.

47. Harriet A. Jacobs, *Incidents in the Life of a Slave Girl: Written by Herself*, original ed. L. Maria Child, ed. Jean Fagan Yellin (Cambridge: Harvard University Press, 1987), 48; John d'Entremont, *Southern Emancipator: Moncure Conway, The American Years 1832-1865* (New York: Oxford University Press, 1987), 21; R. J. M. Blackett, *Beating Against the Barriers: Biographical Essays in Nineteenth-Century Afro-American History* (Baton Rouge: Louisiana State University Press, 1986), 185-86; Lander, *Calhoun Family*, 69-70; Gutman, *Black Family*, 78; Rawick, ed., *American Slave, Texas*, 4: part 1, 240; Livermore, *Story of My Life*, 329-30; Moses Roper, *A Narrative of the Adventures and Escape of Moses Roper from American Slavery*, preface by T. Price (London: Darton, Harvey and Darton, 1838; repr. New York: Negro Universities Press, division of Greenwood Press, 1970), 3; Orlando Patterson, *Slavery and Social Death: A Comparative Study* (Cambridge, Mass.: Harvard University Press, 1982), 6.

48. Rawick, ed., *American Slave, Texas*, 4: part 2, 177-78. Cheryll Ann Cody, "Naming, Kinship, and Estate Dispersal: Notes on Slave Family Life on a South Carolina Plantation, 1786 to 1833," *William and Mary Quarterly* 39 (January 1982): 192-211, discovers that slaves named children for male relatives who were sent to other plantations within South Carolina when their master in the Gaillard family divided his estate in 1825.

49. Philip P. Dandridge to Mrs. R. M. T. Hunter, 15 April 1860, R. M. T. Hunter Papers, UVA; J. A. Townes to W. A. Townes, 4 January 1845, Townes Family Papers, USC; Elizabeth B. Ambler to John J. Ambler, Jr., 10 November 1836, Ambler and Barbour Family Papers, UVA; Will of Ann D. Macrae, Fairfax County Will Book, Book Z-1, 85, 7 June 1858, VSA. See also John C. Inscoe, *Mountain Masters, Slavery, and the Sectional Crisis in Western North Carolina* (Knoxville: University of Tennessee Press, 1989), 91-92; Lebsock, *Free Women of Petersburg*, 137; L. Minor Blackford, *Mine Eyes Have Seen the Glory: The Story of a Virginia Lady Mary Berkeley Minor Blackford 1802-1896 Who Taught Her Sons to Hate Slavery and to Love the Union* (Cambridge, Mass.: Harvard University Press, 1954), 39-40. Cf. Genovese, *Roll, Jordan, Roll*, 452-58, who argues that both planter men and women were aware of the anguish of slave families separated by sales. Patterson, *Slavery and Social Death*, 208, believes that planter women usually understood the oppression of slaves better than planter men because they had "more time and leisure to reflect on" the condition of slaves and perhaps because of their own powerlessness in a "timocratic" culture.

50. Samuel Van Wyck to Margaret Van Wyck, 31 August 1859, Maverick and Van Wyck Family Papers, USC; Philip P. Dandridge to Mrs. R. M. T. Hunter, 29 May 1859, R. M. T. Hunter Papers, UVA; Rawick, ed., *American Slave, Texas*, 4: part 1, 115.

51. Paul W. Gates, *The Farmer's Age: Agriculture 1815-1860*, Economic History of the United States, vol. 3 (New York: Holt, Rinehart and Winston, 1960), 107-9; Smith, *South Carolina*, 54-58; Rosser Howard Taylor, *Slaveholding in North Carolina: An Economic View* (Chapel Hill: University of North Carolina Press,

1926), 30–68; Guion Griffis Johnson, *Ante-bellum North Carolina: A Social History* (Chapel Hill: University of North Carolina Press, 1937), 52–63, 468–92; Craven, *Soil Exhaustion*, 145–46, 122–26, 160–61; McClelland and Zeckhauser, *Demographic Dimensions*, 51, 141–43; James D. Foust, *The Yeoman Farmer and Westward Expansion of United States Cotton Production* (New York: Arno Press, 1975), 73–74, 15–16; Barnes F. Lathrop, *Migration into East Texas 1835–1860: A Study from the United States Census* (Austin: Texas State Historical Association, 1949), 25. McClelland and Zeckhauser show that Virginia, Delaware, Maryland, and the Carolinas (defined as the "Old South") lost an estimated 343,000 whites over the age of ten between 1830 and 1840, including some 55,000 males between the ages of twenty and thirty; they lost an estimated 167,000 between 1840 and 1850, including 27,000 males between twenty and thirty; and they lost 160,000 between 1850 and 1860, including 34,000 males between twenty and thirty. See *Demographic Dimensions*, 141–43.

52. Schlissel, *Women's Diaries*, 14, uses the concept of "antimythic" migration. Theodore W. Schultz, "Migration: An Economist's View," in *Human Migration*, 377, discusses involuntary migrants as a general category.

Chapter 3. A New World: Journey and Settlement

1. John W. DuBose, "A Memoir of Four Families," 2:195, John Witherspoon DuBose Papers, ALA; *Memoirs of Mary A. Maverick, Arranged by Mary A. Maverick and Her Son George Madison Maverick*, ed. Rena Maverick Green (San Antonio: Alamo Printing Company, 1921), 12; Henry E. Blair to Jane I. Blair, 9 October 1849, Blair Papers, VHS. For simplicity's sake, hereafter I will refer to planters' sons who wanted to become planters as "planter men," although not all of them remained in the planter class; I will refer to their female relatives as "planter women." See Chapter 4.

2. Abner J. Grigsby to Reuben Grigsby, 31 October 1846, Grigsby Family Papers, VHS; Basil Hall, *Travels in North America in the Years 1827 and 1828*, 3 vols. (Edinburgh: Cadell and Co., 1829), 3:133; Mary E. P. Allen to John J. Allen, Jr., 28 December 1854, Allen Family Papers, UVA; Sarah H. Brown to Allen Brown, 22 February 1853, Hamilton Brown Papers, UNC.

3. Mary Ann Taylor to Marion Singleton, 2 April 1832, Singleton Family Papers, LC; *The Lides Go South . . . And West: The Records of a Planter Migration in 1835*, ed. Fletcher M. Green (Columbia: University of South Carolina Press, 1952), 8, n. 19; Thomas Felix Hickerson, *Happy Valley: History and Genealogy* (Chapel Hill: by the author, 1940), 113. On the sadness of other American women leaving for other frontiers, see Annette Kolodny, *The Land Before Her: Fantasy and Experience of the American Frontiers* (Chapel Hill: University of North Carolina Press, 1984), 94–95; Lillian Schlissel, *Women's Diaries of the Westward Journey* (New York: Schocken Books, 1982), 14, 28; Joanna L. Stratton, *Pioneer Women: Voices from the Kansas Frontier* (New York: Simon and Schuster, 1981), 34; Julie Roy Jeffrey, *Frontier Women: The Trans-Mississippi West, 1840–1880*, American Century Series (New York: Hill and Wang, 1979), 37.

4. *The American Slave: A Composite Autobiography*, ed. George P. Rawick,

Contributions in Afro-American and African Studies, No. 11 (Westport, Conn.: Greenwood Publishing Company, 1972), *Alabama*, 6:72; DuBose, "Memoir," 2:203, John Witherspoon DuBose Papers, ALA; Arthur Calhoun, *A Social History of The American Family, From Colonial Times to the Present*, 3 vols. (Cleveland: Arthur H. Clark Company, 1918; repr. New York: Arno Press, 1973), 2:269. Binah DuBose met her son about twenty years later when her own master, Kimbrough DuBose, migrated to Alabama.

5. DuBose, "Memoir," 2:194–95, John Witherspoon DuBose Papers, ALA. See also Rawick, ed., *American Slave, South Carolina*, 2: part 1, 40; *Alabama*, 6:413–16; Norman R. Yetman, *Life Under the 'Peculiar Institution': Selections from the Slave Narrative Collection* (Huntington, N.Y.: Robert E. Krieger, 1976), 145. During and after the Civil War, thousands of blacks traveled the South seeking relatives they had lost through westward migration in the antebellum years; see Leon F. Litwack, *Been in the Storm So Long: The Aftermath of Slavery* (New York: Alfred A. Knopf, 1980).

6. Rhys Isaac, *The Transformation of Virginia, 1740–1790* (Chapel Hill: published for Institute of Early American History and Culture by University of North Carolina Press, 1982), 52–57, notes that slaves, poor whites, planter whites, white men, and white women experienced the eighteenth-century Virginian landscape in diverse ways.

7. William O. Lynch, "The Westward Flow of Southern Colonists before 1861," *Journal of Southern History* 9 (1943): 314; Peter D. McClelland and Richard J. Zeckhauser, *Demographic Dimensions of the New Republic: American Interregional Migration, Vital Statistics, and Manumissions, 1800–1860* (Cambridge: Cambridge University Press, 1982), 51. Nineteen men in Table 2 migrated in the 1830s; thirteen migrated in the 1840s; seven in the 1850s; four in the 1820s; three in the 1810s; there is no information at all for three men. McClelland and Zeckhauser, *Demographic Dimensions*, also argue (6, 51) that the "New South" (defined as Georgia, Florida, Alabama, Mississippi, Tennessee, Louisiana, and Arkansas) received the largest influx of white migrants between 1810 and 1820; many of those who moved into the New South in this decade must have been from regions other than the Old South. They exclude Texas, the destination of many Southerners between 1850 and 1860, from their calculations.

8. Edmund Jones Diary, vol. 32, n.d. 1837, Jones and Patterson Papers, UNC. Thomas Brown, "An Account of the Lineage of the Brown Family," 60, Ambler-Brown Family Papers, DU; John Mack Faragher, *Sugar Creek: Life on the Illinois Prairie* (New Haven: Yale University Press, 1986), 57; idem, *Women and Men on the Overland Trail* (New Haven: Yale University Press, 1979), 33, notes that kinfolk accompanied travelers on the western trail.

9. On departures and travel, see F. J. Krone to William Gaston, 5 January 1834, William Gaston Papers, UNC; Edmund Jones Diary, vol. 32, n.d. 1837, Jones and Patterson Papers, UNC. On routes, see John E. Dromgoole to Edward Dromgoole, Sr., 28 January 1832, Edward Dromgoole Papers, UNC; S. A. Townes to G. F. Townes, 26 August 1834, Townes Family Papers, USC; Henry DeLeon Southerland, Jr., "The Federal Road, Gateway to Alabama," *The Alabama Review: A Quarterly Journal of Alabama History* 34 (April 1986): 96, 98; Geological Survey,

United States Department of the Interior, *The National Atlas of the United States of America* (Washington, D.C., 1970), 136-38.

10. Rawick, ed., *American Slave, South Carolina,* 3: part 3, 75; Elisabeth Muhlenfeld, *Mary Boykin Chesnut: A Biography,* Southern Biography Series, ed. William J. Cooper, Jr. (Baton Rouge: Louisiana State University Press, 1981), 30; Harriet Martineau, *Society in America,* 3 vols. (London: Saunders and Otley, Conduit Street, 1837; repr. New York: AMS Press, 1966), 1:293. Of the forty-nine migrants in Table 2, twenty-three came from North Carolina and thirteen each from Virginia and South Carolina. Twenty men migrated to Alabama; eleven men to Tennessee; ten to Mississippi; four to Louisiana; and one each to Arkansas, California, Florida, and Texas. According to Richard H. Steckel's calculations based on the *1850* census, the choice of Alabama as a destination is not statistically representative of the Virginians, North Carolinians, and South Carolinians who lived outside of their states of birth in 1850. Among Virginian migrants, most went to Ohio—22.1 percent—and only 2.7 percent went to Alabama. Among North Carolinians, most went to Tennessee—25.4 percent—and only 10.1 percent moved to Alabama. Among South Carolinians, 28 percent went to Georgia, and 26.1 percent lived in Alabama in 1850. See "The Economic Foundations of East-West Migration during the 19th Century," *Explorations in Economic History* 20 (1983): 14-36.

11. On the length of the journey, see Israel Pickens to William B. Lenoir, 2 June 1823, Chiliab Smith Howe Papers, UNC; Diary attributed to Kiziah Hardy, November 1855, 50, S. H. Steelman Collection, UNC; Green, ed., *Lides,* Journal of Sarah Jane (Lide) Fountain, November 18-December 26, 1835, 3-8; Daniel M. Johnson and Rex R. Campbell, *Black Migration in America: A Social Demographic History* (Durham, N.C.: Duke University Press, 1981), 30. On its cost, see S. A. Townes to Rachel Townes, 13 November 1833, Townes Family Papers, USC; Eli P. Whitaker to Henry Whitaker, 12 May 1835, Whitaker and Snipes Family Papers, UNC; Charles Cooke to Callendar Noland, 4 June 1857, Noland Family Papers, UVA. Costs declined over the decades as roads improved so that by the 1850s the trip cost only several hundred dollars. Seaboard migrants to midwestern and transcontinental frontiers traveled for longer periods of time, from six to eight months. See Faragher, *Sugar Creek,* 56; Schlissel, *Women's Diaries,* 24; John D. Unruh, Jr., *The Plains Across: The Overland Emigrants and the Trans-Mississippi West, 1840-1860* (Urbana: University of Illinois Press, 1979), 58-82, 198-251, 341. Journeys on the Overland Trail cost from five hundred to a thousand dollars for one family. See Schlissel, *Women's Diaries,* 23; Jeffrey, *Frontier Women,* 28.

12. Richard C. Ambler to John J. Ambler, Jr., n.d. 1835, Ambler and Barbour Family Papers, UVA; Brown, "Brown Family," 61, Ambler-Brown Family Papers, DU; R. R. Hackett to R. F. Hackett, 31 October 1848, Gordon and Hackett Family Papers, UNC. Faragher, *Overland Trail,* 89-104, 135-36, shows that men on the Overland Trail perceived the journey as an affirmation of masculinity.

13. Rawick, ed., *American Slave, South Carolina,* 3: part 3, 75; Frederick Law Olmsted, *The Cotton Kingdom: A Traveller's Observations on Cotton and Slavery in the American Slave States,* ed. Arthur M. Schlesinger, Sr., introduction by Lawrence N. Powell (New York: Modern Library, 1984), 285; Muhlenfeld, *Chesnut,* 28-29.

14. Green, ed., *Lides,* Journal of Sarah Jane (Lide) Fountain, 20 November 1835, 1; Rawick, ed., *American Slave, South Carolina,* 3: part 3, 75; Elizabeth Hord to Martha S. Burrus, 13 November 1835, Burrus Family Papers, VHS; Claude Kitchin to A. M. Arnett, 25 November 1935, Kitchin Papers, UNC; Joseph H. Parks, *General Leonidas Polk C.S.A., the Fighting Bishop* (Baton Rouge: Louisiana State University Press, 1962), 65. Most scholars note that the trip was especially difficult for women; see Kolodny, *Land Before Her,* 31–34; Schlissel, *Women's Diaries,* 14–16, 95, 111–12, 115; Jeffrey, *Frontier Women,* 38–50. Faragher, *Overland Trail,* 75–87, argues that women did more arduous work and worked longer hours than men did; women also worked harder than they had at home. Cf. Glenda Riley, *Frontierswomen: The Iowa Experience* (Ames: Iowa State University Press, 1981), 28, and Sandra Myres, *Westering Women and the Frontier Experience, 1800–1915* (Albuquerque: University of New Mexico Press, 1982), 98–139, who emphasize the ease with which women coped with these difficulties.

15. Narrative of Jane Harris Woodruff, 311–17, Harris Papers, SCHS.

16. Rawick, ed., *American Slave, Mississippi,* 7:151–52; *Tennessee,* 16:76; *Alabama,* 6:58; William Whitfield to Gaius Whitfield, 2 March 1837, Gaius Whitfield Papers, ALA; Leslie Howard Owens, *This Species of Property: Slave Life and Culture in the Old South* (New York: Oxford University Press, 1976), 190; Rawick, ed., *Alabama,* 6:48, *Texas,* 5: part 4, 214; Hall, *Travels in North America,* 3: 129; Sydney Nathans, "Fortress without Walls: A Black Community after Slavery," in *Holding on to the Land and the Lord: Kinship, Ritual, Land Tenure, and Social Policy in the Rural South,* ed. Robert L. Hall and Carol B. Stack, Southern Anthropological Society Proceedings, No. 15, Robert L. Blakely, series editor (Athens: University of Georgia Press, 1982), 56; Johnson and Campbell, *Black Migration,* 29–30. For an account of a slave who ran away from his Mississippi master intending to go to South Carolina, see James M. Wylie to "Mr. Dixon," 12 August 1855, Gaston, Strait, Wylie, and Baskin Families Papers, USC. Some masters sent their slaves down the Ohio or Mississippi rivers rather than by overland march; see Frederic Bancroft, *Slave-trading in the Old South* (Baltimore: J. H. Furst Company, 1931), 275–81.

17. Hall, *Travels in North America,* 3:127–28; George Lee Simpson, Jr., *The Cokers of Carolina: A Social Biography of a Family* (Chapel Hill: Published for Institute for Research in Social Science by University of North Carolina Press, 1956), 41; Virginia Clay-Clopton, *A Belle of the Fifties: Memoirs of Mrs. Clay of Alabama, Covering Social and Political Life in Washington and the South, 1853–1866, put into narrative form by Ada Sterling* (New York: Doubleday, Page & Company, 1905), 5; Ruth Ketring Nuermberger, *The Clays of Alabama: A Planter-Lawyer-Politician Family* (Lexington: University of Kentucky Press, 1958), 82; *From Virginia to Texas, 1835: Diary of Colonel William F. Gray Giving Details of His Journey to Texas and Return in 1835–1836 and Second Journey to Texas in 1837,* preface by A. C. Gray (Houston: Gray, Dillaye & Co., 1909; repr. Houston: Fletcher Young Publishing Co., 1965), vii; Richard C. Ambler to John J. Ambler, Jr., n.d. 1835, Ambler and Barbour Family Papers, UVA.

18. N. Boykin to John Cantey, 19 August 1834, Cantey Family Papers, USC; Edmund Jones Diary, vol. 32, n.d. 1837, Jones and Patterson Papers, UNC; M. H.

Martin to Benjamin C. Yancey, 10 March 1837, Benjamin Cudworth Yancey Papers, UNC; Thomas Barbour to John J. Ambler, Jr., and Elizabeth B. Ambler, 26 January 1834, Ambler and Barbour Family Papers, UVA.

19. Henry A. Tayloe to Benjamin O. Tayloe, 26 December 1833, Tayloe Family Papers, UVA; M. C. Stephens to William Gaston, 19 June 1840, William Gaston Papers, UNC; Billy Mac Jones, *Health Seekers in the Southwest, 1817–1900* (Norman: University of Oklahoma Press, 1967), 3–10, 14–22, 28, 35; Abner and G. H. Grigsby to Reuben Grigsby, 15 January 1846, Grigsby Family Papers, VHS; William Connell to J. Turner and C. C. Cooper, n.d. 1853, Odum-Turner Family Papers, USC; Samuel A. Stouffer, "Intervening Opportunities: A Theory Relating Mobility and Distance," *American Sociological Review* 5 (December 1940): 846. Yeomen generally preceded planters to the Southwest and settled the upland regions rather than the alluvial lowlands; see John Solomon Otto, "The Migration of the Southern Plain Folk: An Interdisciplinary Synthesis," *Journal of Southern History* 51 (May 1985): 193–94. Most men in Table 2 went directly from their homes in the seaboard to the Southwest rather than moving westward in stages across the seaboard.

20. Obituary of William P. Duval, *Hinds County (Miss.) Gazette*, 7 June 1854; Hugh Thomas Brown to Caroline Gordon Hackett, 11 June 1860, 2 August 1860, Gordon and Hackett Family Papers, UNC; Christopher C. Scott Autobiography, 1848, 12, 13, 22, Christopher C. Scott Family Notes, UNC; S. A. and Joanna Townes to G. F. Townes, 10 March 1835, S. A. and Joanna Townes to J. A. Townes, 18 September 1834, Townes Family Papers, USC. Robert E. Bieder, "Kinship as a Factor in Migration," *Journal of Marriage and the Family* 35 (August 1973): 429–39, argues that kinship was crucial in recruiting migrants to a mid-nineteenth century Michigan community and in reducing outmigration.

21. R. R. Hackett to R. F. Hackett, 14 September 1849, Gordon and Hackett Family Papers, UNC; Josiah C. Nott to James M. Gage, 28 July 1836, James McKibbon Gage Papers, UNC; Littleton Turner to John Turner, 7 July 1837, Odum-Turner Family Papers, USC.

22. Obituary of Elizabeth A. Cully, *Hinds County (Miss.) Gazette*, 26 July 1854; Hickerson, *Happy Valley*, 78. See also Gaius Whitfield to Needham Whitfield, 2 July 1835, Whitfield and Wooten Family Papers, UNC; Catherine Clinton, *The Plantation Mistress: Woman's World in the Old South* (New York: Pantheon Books, 1982), 166–67.

23. H. Vaughan to John B. Miller, 12 December 1836, Miller-Furman-Dabbs Papers, USC; Gaius Whitfield to Needham Whitfield, 2 July 1835, James B. Whitfield to Needham Whitfield, 30 September, 1837, 3 October 1837, Whitfield and Wooten Family Papers, UNC. Many aspects of the pioneer phase, especially the hard work involved, invite comparison with the colonial era when the Southern seaboard was settled.

24. Paul W. Gates, *The Farmer's Age: Agriculture 1815–1860*, Economic History of the United States, vol. 3 (New York: Holt, Rinehart and Winston, 1960), 140–41; Lewis Cecil Gray, *History of the Agriculture of the Southern United States to 1860*, 2 vols. (Baltimore: Waverly Press, Inc., 1933), 2:895, 642–44; Mead Carr to Bernard Carr, 3 December 1831, George Carr Papers, UVA; Gaius Whitfield to Needham Whitfield, 30 July 1835, Whitfield and Wooten Family Papers, UNC.

25. Gray, *Agriculture*, 2:1027, 899, 666; Account Book, n.p.; "H. A. Tayloe's Liabilities, 23 October 1843, Estimate value of the Ala. plantations," Tayloe Family Papers, VHS; Henry A. Tayloe to Benjamin O. Tayloe, n.d. 1836, Tayloe Family Papers, UVA; Ernest McPherson Lander, Jr., *The Calhoun Family and Thomas Green Clemson: The Decline of a Southern Patriarchy* (Columbia: University of South Carolina Press, 1983), 36; Rawick, ed., *American Slave, Mississippi*, 7:157; Gaius Whitfield to Needham Whitfield, 2 July 1835, Whitfield and Wooten Family Papers, UNC. In the wake of Nat Turner's rebellion in Virginia in 1831, Mississippi prohibited the importation of slaves for the purposes of sale (as opposed to slaves who came west with their owners) from 1833 to 1837, when the law was repealed. It does not seem to have been enforced, however, since the state simultaneously taxed such slave imports. See Wendell Holmes Stephenson, *Isaac Franklin: Slave Trader and Planter of the Old South, with Plantation Records* (Baton Rouge: Louisiana State University Press, 1938), 61-62. The average acreage per farm (not per plantation) in the Southwest in *1850* was 273 acres, according to Gray, *Agriculture*, 1:530.

26. Lander, *Calhoun Family*, 7-12, 22-26; Journal of William A. Lenoir, 3 November 1836, vol. 155, Lenoir Family Papers, UNC; Henry A. Tayloe to Benjamin O. Tayloe, 14 February 1836, Tayloe Family Papers, UVA.

27. Gray, *Agriculture*, 2: 898-900; Larry Schweikart, *Banking in the American South from the Age of Jackson to Reconstruction* (Baton Rouge: Louisiana State University Press, 1987), 59; Peter Temin, *The Jacksonian Economy* (W. W. Norton & Company, 1969), 90-91; Davis R. Dewey, *State Banking before the Civil War* (Washington, D.C.: Government Printing Office, 1910), 160, 163; S. A. Townes to G. F. Townes, 24 August 1834, Townes Family Papers, USC. For a satire on the buying frenzy of the 1830s, see Joseph G. Baldwin, *The Flush Times of Alabama and Mississippi: A Series of Sketches* (New York: D. Appleton and Company, 1853), 81-84.

28. Josiah C. Nott to James M. Gage, 28 July 1836, James McKibbon Gage Papers, UNC; Henry A. Tayloe to Benjamin O. Tayloe, 6 February 1836, Tayloe Family Papers, UVA; Hickerson, *Happy Valley*, 117-18.

29. *Diary of William Gray*, 28; Hickerson, *Happy Valley*, 78; Green, ed., *Lides*, 17; Richard C. Ambler to John J. Ambler, Jr., 11 December 1835, Ambler and Barbour Family Papers, UVA.

30. Green, ed., *Lides*, 17; Lewis E. Caperton to Henry Caperton, 12 March 1853, Caperton Family Papers, VHS; S. A. Townes to G. F. Townes, 7 October 1834, Townes Family Papers, USC; Josiah C. Nott to James M. Gage, 28 July 1836, James McKibbon Gage Papers, UNC. Other American men in the early nineteenth century thought land beautiful if it was cleared, well-cultivated, and capable of supporting a family; see John R. Stilgoe, *Common Landscape of America, 1580-1845* (New Haven: Yale University Press, 1982), 206.

31. Hickerson, *Happy Valley*, 116-17; Israel Pickens to William B. Lenoir, 18 January 1818, Israel Pickens Papers, ALA; S. A. Townes to John A. Townes, 1 October 1834, S. A. Townes and Joanna Townes to J. A. Townes, 18 September 1834, S. A. Townes to G. F. Townes, 8 September 1834, 27 February 1834, 24 August 1834, 14 September 1835, Townes Family Papers, USC.

32. O. G. Murrell to John Murrell, 11 October 1849, Murrell Family Papers, UVA; Richard T. Archer to Stephen Archer, 18 March 1833, Archer Papers, UT; "Martha" to her mother, 1 January 1858, Scarborough Family Papers, DU.

33. John J. Ambler, Jr., to Elizabeth B. Ambler, 25 October 1836, Ambler and Barbour Family Papers, UVA; *The Papers of John C. Calhoun*, Volume 15, *1839–1841*, ed. Clyde N. Wilson (Columbia: published by University of South Carolina Press for South Carolina Department of Archives and History and South Caroliniana Society, 1983); 319, Andrew Pickens Calhoun to John C. Calhoun, 2 August 1840; S. A. Townes to J. A. Townes, n.d. August 1836, Townes Family Papers, USC; William S. Brown to Henry J. Brown, 19 February 1852, Henry James Brown Papers, UVA. On climate and illness, see James H. Cassedy, "Medical Men and the Ecology of the Old South," in *Science and Medicine in the Old South*, ed. Ronald L. Numbers and Todd L. Savitt (Baton Rouge: Louisiana State University Press, 1989), 168, 173; Sam Bowers Hilliard, *Atlas of Antebellum Southern Agriculture*, cartography by Clifford P. Duplechin and Mary Lee Eggart (Baton Rouge: Louisiana State University Press, 1984), 14–15; Lucie Robertson Bridgforth, "Medicine in Antebellum Mississippi," *Journal of Mississippi History* 46 (May 1984): 82–107; Charles E. Rosenberg, *The Cholera Years: The United States in 1832, 1849, and 1866* (Chicago: University of Chicago Press, 1962), 3–4, 36–37, 60–61, 105, 115–16; Todd L. Savitt, *Medicine and Slavery: The Diseases and Health Care of Blacks in Antebellum Virginia*, Blacks in the New World, August Meier, series editor (Urbana: University of Illinois Press, 1978), 21–26, 34.

34. Henry A. Tayloe to B. Ogle Tayloe, 7 January 1834, 5 January 1835, 28 January 1837, Tayloe Family Papers, UVA; Brown, "Brown Family," 61–62, Ambler-Brown Family Papers, DU; R. R. Hackett to R. F. Hackett, 15 November 1851, Gordon and Hackett Family Papers, UNC.

35. Benjamin Scarborough to Samuel Scarborough, 27 September 1840, 24 April 1842, Scarborough Family Papers, DU; Willis Lea to William Lea, 15 November 1841, Lea Family Papers, UNC; Louisa Cunningham to Benjamin Yancey, 30 August 1838, Benjamin Cudworth Yancey Papers, UNC.

36. James R. Deupree to William Deupree, 22 December 1839, William Deupree Papers, VHS; W. F. Hunt to Benjamin F. Perry, 25 March 1857, Benjamin Franklin Perry Papers, UNC; Samuel Van Wyck to Margaret B. Van Wyck, 23 March 1860, Maverick and Van Wyck Family Papers, USC.

37. Mary Ann Taylor to Marion Singleton, n.d. January 1834, Singleton Papers, LC; L. A. Gray to R. F. Hackett, 11 January 1858, Gordon and Hackett Family Papers, UNC; "Martha" to her mother, 1 January 1858, Scarborough Family Papers, DU.

38. M. E. A. Drake to Gray and Louisa Sills, 14 June 1839, Jelks Sills Letters, DU; Elisabeth Showalter Muhlenfeld, "Mary Boykin Chesnut: The Writer and Her Work" (Ph.D. dissertation, University of South Carolina, 1978), Appendix C, 824; "Memoir of Harriet M. P. P. Ames," 5, Potter Papers, DU; Caroline L. Gordon to her brother, 20 May 1853, James Gordon Hackett Papers, DU; Drake to Sills, 14 June 1839, Jelks Sills Letters, DU. William Cronon, *Changes in the Land: Indians, Colonists, and the Ecology of New England* (New York: Hill and Wang, 1983), shows how Indians and Britons perceived the New England landscape in pro-

foundly different ways. See *Humanscape: Environments for People,* ed. Stephen Kaplan and Rachel Kaplan (Ann Arbor: Ulrich's Books, 1982), for other work by geographers on how individuals perceive landscapes.

39. Adelaide Crain to Caroline Gordon, 3 April 1857, Gordon and Hackett Family Papers, UNC; Daniel and Mary Kelly to James C. and Margaret Kelly, 29 February 1845, John N. Kelly Papers, DU: Lewis E. Caperton to Henry Caperton, 22 March 1858, Caperton Family Papers, VHS. Myres, *Westering Women,* 138–39, 156, and Stratton, *Pioneer Women,* 71, describe women's health problems on other frontiers.

40. Sterling Graydon to Mr. and Mrs. Mabra Madden, 10 August 1856, Mabra Madden Papers, USC. On the work women performed on other frontiers, see Myres, *Westering Women,* 141–42; Riley, *Iowa,* 36–37; Jeffrey, *Frontier Women,* 53. Robert F. Berkhofer, Jr., "Space, Time, Culture, and the New Frontier," in *Geographical Perspectives on America's Past,* ed. David Ward, with assistance of Thomas S. Flory (New York: Oxford University Press, 1979), 44, notes that the cultural values settlers bring with them strongly affect how they respond to an environment.

41. On housing, see "Personal Reflections," 2, Recollections of Frances Rebecca Bouldin Spragins Brown, VHS; Richard C. Ambler to John J. Ambler, Jr., 11 December 1835, Ambler and Barbour Family Papers, UVA; Emma Morehead Whitfield, *Whitfield, Bryan, Smith, and Related Families,* ed. Theodore Marshall Whitfield, 2 vols, (n.p., 1948), 1:90; Adelaide Crain to Caroline Gordon, 3 September 1857, Gordon and Hackett Family Papers, UNC. For quoted sources, see Muhlenfeld, "Chesnut," 3:579; Rawick, ed., *American Slave, South Carolina,* 3: part 3, 76; Edward F. Wells to Nora and Tommy Wells, 26 February 1856, Wells Family Papers, UVA. Regarding Mary Chesnut's account of her family's migration, C. Vann Woodward concludes that it is "clearly fiction" but "autobiographical in substance." See *Mary Chesnut's Civil War,* ed. C. Vann Woodward (New Haven: Yale University Press, 1981), xxiii.

42. Sterling Graydon to Mr. and Mrs. Mabra Madden, 10 August 1856, Mabra Madden Papers, USC; Muhlenfeld, *Chesnut,* 17, 30; H. B. Eggleston to "My dear Cousin," 13 April 1859, Archer Papers, UT.

43. "Memoir of Ames," 10, Potter Papers, DU; Marcus C. Stephens to William Gaston, 28 May 1831, William Gaston Papers, UNC. On food and clothing, see Nancy Gillespie to Margaret Murphey, 8 September 1827, Neill Brown Papers, DU; J. N. Berryman to Leroy H. Berryman, 30 July 1834, Berryman Family Papers, VHS; John Micou, Jr., to R. M. T. Hunter, 21 April 1836, R. M. T. Hunter Papers, UVA; M. E. A. Drake to Gray and Louisa Sills, 14 June 1839, Jelks Sills Letters, DU: S. A. Townes to Rachel Townes, 2 November 1834, Townes Family Papers, USC. Stratton, *Pioneer Women,* 62–65, and Jeffrey, *Frontier Women,* 54, mention the difficulties of food preparation in other western settlements.

44. Claudia Dale Goldin, *Urban Slavery in the American South, 1820–1860: A Quantitative History* (Chicago: University of Chicago Press, 1976), 64–66.

45. Henry Tayloe to Benjamin O. Tayloe, 29 July 1839, Tayloe Family Papers, UVA; James R. Creecy, *Scenes in the South, and other Miscellaneous Pieces* (Washington, D.C.: Thomas McGill, 1860), 83; Richard C. Ambler to John J. Ambler, Jr., 11 December 1835, 15 January 1836, Ambler and Barbour Family

Papers, UVA; S. A. Townes to Rachel Townes, 2 November 1834, Townes Family Papers, USC; Narrative of Jane Harris Woodruff, 325, Harris Papers, SCHS.

46. S. A. Finley to her sister, 21 December 1854, S. A. Finley to Sarah Brown, 23 November 1855, S. A. Finley to Caroline Gordon Hackett, 8 August 1860, Gordon and Hackett Family Papers, UNC; Ann Archer to Abram Archer, 8 March 1857, Archer Papers, UT; Charles Cooke to "Dear Cousin Call," 4 June 1857, Noland Family Papers, UVA. Susan Strasser, *Never Done: A History of American Housework* (New York: Pantheon Books, 1982), 32–49, 50–66, describes how technological advancements in the mid-nineteenth century, such as the iron stove, made housework easier for all American women.

47. Myres, *Westering Women*, 160; Stratton, *Pioneer Women*, 61.

48. Woodruff Narrative, 317–23, Harris Papers, SCHS; Minnie Clare Boyd, *Alabama in the Fifties: A Social Study*, Studies in History, Economics and Public Law, ed. Faculty of Political Science of Columbia University (New York: Columbia University, 1931), 45; S. A. Townes to J. A. Townes, n.d. August 1836, Townes Family Papers, USC; Woodruff Narrative, 322–32, Harris Papers, SCHS; Ray Mathis, *John Horry Dent: South Carolina Aristocrat on the Alabama Frontier* (University: published under the sponsorship of Historic Chattahoochee Commission by University of Alabama Press, 1979), 158.

49. S. A. Finley to Caroline L. Gordon, 17 February 1853, Gordon and Hackett Family Papers, UNC; S. A. Townes to G. F. Townes, 7 October 1834, Townes Family Papers, USC; Caroline L. Gordon to her brother, 20 May 1853, James Gordon Hackett Papers, DU; W. A. Townes to J. A. Townes, 14 January 1844, Townes Family Papers, USC. See Israel Pickens to William B. Lenoir, 25 November 1823, Chiliab Smith Howe Papers, UNC, for one master who recognized that slaves needed a few years to become acclimated to the Southwest. Savitt, *Medicine and Slavery*, is an indispensable source on the history of frontier medicine and diseases; he argues (7–47) that many antebellum whites believed blacks had special immunities and susceptibilities to various diseases. Many blacks could in fact resist variants of malaria better than whites because of the immunity provided by sickle cell anemia, but they were more susceptible to pulmonary infections. They were also more resistant to yellow fever, for reasons that medical scientists still do not understand (240–46), and they were somewhat better at tolerating hot, humid weather because they released fewer body salts in sweat and urine (40–41).

50. Allen Brown to Hamilton Brown, 8 December 1841, M. Brown to Sarah Brown, 15 January 1842, Hamilton Brown Papers, UNC.

51. Thomas and Anna Meade to Harriet E. Meade, 18 March 1849, Whitaker and Meade Family Papers, UNC; Thomas Felix Hickerson, *Echoes of Happy Valley: Letters and Diaries, Family Life in the South, Civil War History* (Chapel Hill: by the author, 1962), 6; Thomas Tabb to Allen A. Burwell, 8 March 1832, Roberts Family Papers, VHS; G. F. Townes to J. A. Townes, 4 February 1834, S. A. Townes to G. F. Townes, 16 February 1834, Townes Family Papers, USC.

52. Caroline Gordon to H. Thomas Brown, 13 November 1854, Caroline Gordon to H. Thomas Brown, 24 February 1852, Gordon and Hackett Family Papers, UNC; Green, ed., *Lides*, 19.

53. Woodruff Narrative, 317, Harris Papers, SCHS. Herbert Gutman, *The Black Family in Slavery and Freedom, 1750–1925* (New York: Pantheon Books, 1976), 155–65, argues that frontier slaves eventually re-created the values and "essential domestic arrangements" of older slave communities.

54. Philip St. George Ambler to John J. Ambler, Jr., 5 January 1836, Richard C. Ambler to John J. Ambler, Jr., 6 February 1836, Elizabeth B. Ambler to John J. Ambler, Jr., 20 June 1836, John J. Ambler, Jr., to Elizabeth B. Ambler, 17 October 1836, 25 October 1836, 8 November 1836, Richard C. Ambler to John J. Ambler, Jr., 17 November 1837, John J. Ambler, Jr., "Memoranda," 296, Ambler and Barbour Family Papers, UVA. Indians sporadically raided white settlements long after they had been defeated in the Creek War in 1814; see Daniel H. Usner, "American Indians on the Cotton Frontier: Changing Economic Relations with Citizens and Slaves in the Mississippi Territory," *Journal of American History* 72 (September 1985): 315–17.

55. William S. Brown to Henry J. Brown, 19 February 1852, Henry James Brown Papers, UVA; Duncan McKenzie to Duncan McLaurin, n.d. July 1840, Duncan McLaurin Papers, DU. Michael Conzen, "Local Migration Systems in Nineteenth-Century Iowa," *Geographical Review* 64 (July 1974): 339, notes the frequent return migration from all American frontiers. Donald H. Parkerson, "How Mobile were Nineteenth-Century Americans?" *Historical Methods: A Journal of Quantitative and Interdisciplinary History* 15 (Summer 1982): 105, observes that persistence could mean that settlers were trapped. Seven of the forty-nine migrants in Table 2, or 14 percent, returned to the seaboard.

56. Philip St. George Ambler to John J. Ambler, Jr., 5 January 1836, Elizabeth B. Ambler to John J. Ambler, Jr., 20 June 1836, John J. Ambler, Jr. to Elizabeth B. Ambler, 17 October 1836, 25 October 1836, 8 November 1836, Richard C. Ambler to John J. Ambler, Jr., 17 November 1837, Ambler and Barbour Family Papers, UVA; Lander, *Calhoun Family*, 107–10, 117, 132–35, 142–49, 222–23, 233; Caroline Gordon to H. Thomas Brown, 5 March 1855, Gordon and Hackett Family Papers, UNC.

57. Eustace C. Moncure, Jr., to Eustace C. Moncure, Sr., 23 October 1857, Eustace Conway Moncure Papers, VHS; Biographical fragment on Benjamin Yancey, n.p., n.d., Benjamin Cudworth Yancey Papers, UNC.

58. John J. Ambler, Jr. to Elizabeth B. Ambler, 13 September 1846, 24 January 1848, 30 May 1851, 4 June 1851, Elizabeth B. Ambler to John J. Ambler, Jr., 26 September 1846, Ambler and Barbour Family Papers, UVA; Lewis E. Caperton to Henry Caperton, 22 March 1858, Caperton Family Papers, VHS; John B. Dabney Manuscript, 51, UVA.

59. Clarence L. Robards to Louise Hill, 3 July 1854, Daniel S. Hill Papers, DU; Lewis E. Caperton to Henry Caperton, 12 March 1853, Caperton Family Papers, VHS; Muhlenfeld, *Chesnut*, 34–37.

60. Rawick, ed., *American Slave, South Carolina*, 3: part 3, 74–78; Eighth Decennial Census of the United States, Free Schedule, South Carolina, Union County, 271; Rita Jones Elliott, "The Herndon and Connor Families, Kith and Kin," North Carolina State Library, Raleigh. On the homesickness of slaves for the seaboard, see Solomon Northup, *Narrative of Solomon Northup, a Citizen of New York, Kid-*

napped in Washington City in 1841, and Rescued in 1853, from a Cotton Plantation near the Red River in Louisiana (New York: Miller, Orton & Mulligan, 1855), 186.

61. Duncan McKenzie to Duncan McLaurin, 19 February 1840, Duncan McLaurin Papers, DU. On migration, see James E. Rice to William Deupree, 30 May 1846, George Carr Papers, UVA; Thomas Lewers to Hugh Saxon, 17 February 1847, Hugh Saxon Manuscripts, USC; Robert J. Brugger, *Beverly Tucker: Heart over Head in the Old South* (Baltimore: Johns Hopkins University Press, 1978), 72. On motives, see Fenton Noland to William Noland, 28 September 1854, Noland Family Papers, UVA; Shelby Carr to Bernard Carr, 12 May 1846, George Carr Papers, UVA; "Memoir of Ames," 2, Potter Papers, DU; Henry A. Tayloe to [no name], 15 July 1835, Tayloe Family Papers, UVA; Samuel Van Wyck to "Dear Gus," 1 November 1860, Maverick and Van Wyck Family Papers, USC. Terry G. Jordan, "The Imprint of the Upper and Lower South on Mid-Nineteenth-Century Texas," in *Geographical Perspectives*, ed. Ward, 210–26, shows that cotton planters from what he calls the Lower South (Alabama, Mississippi, and Louisiana) clustered in several counties on the Gulf Coast of Texas, while yeoman farmers from the Upper South (Virginia, Kentucky, and Tennessee) tended to go inland and settle interior counties.

62. Thomas Barbour to Elizabeth B. Ambler, 25 May 1836, Ambler and Barbour Family Papers, UVA; O. G. Murrell to John Murrell, 20 October 1843, Murrell Family Papers, UVA; R. H. Hill to Daniel S. Hill, 17 October 1853, Daniel S. Hill Papers, DU; Autobiographical fragment by Duff Green, 25, 27, Duff Green Papers, UNC; "Another Chapter on Texas," *DeBow's Review* 23 (December 1857): 570–75; *Diary of William Gray*, 84, 91.

63. Mathis, *John Horry Dent*, 23, 45–46; copy of letter to Lawrence Witsell, 24 August 1855, Dent Diary, 24 August 1855, 18 June 1856, copy of letter to "Dear Sir," 3 September 1856, Dent Papers, University of Alabama. For unknown reasons, Dent postponed moving until the Civil War broke out; after the war he moved to Georgia, where he died. Cf. John Hebron Moore, *The Emergence of the Cotton Kingdom in the Old Southwest: Mississippi, 1770–1860* (Baton Rouge: Louisiana State University Press, 1988), 18, 132, 287, who argues that many Mississippi planters adopted less destructive agricultural methods after the depression of the late 1830s.

64. S. Jane Boyd to Jane G. Crawford, 8 October 1852, Gaston-Crawford Papers, USC; "Life of Amelia Lewis Thompson," 9–10, Somerville-Howorth Collection, Schlesinger Library, Radcliffe College; Green, ed., *Lides*, 35.

65. S. A. Townes to H. H. Townes, 10 September 1839, Townes Family Papers, USC; Mathis, *John Horry Dent*, 41; M. E. A. Drake to Gray Sills and Louisa Sills, 14 June 1839, Jelks Sills Letters, DU.

66. Gayle Diary, 13 December 1828, n.d. November 1828, Bayne and Gayle Family Papers, UNC; "Sippy" [Virginia Gordon] to Harriet E. Brown, 8 August 1849, Whitaker and Meade Family Papers, UNC; Carney Diary, n.d., spring 1860, UNC. Jeffrey, *Frontier Women*, 63, notes that men on the trans-Mississippi frontier usually decided whether their families would migrate within the West.

67. Henry A. Tayloe to Benjamin O. Tayloe, 21 February 1839, 29 July 1839,

Tayloe Family Papers, UVA; S. A. Townes to G. F. Townes, 22 June 1834, Townes Family Papers, USC. On the separation of slave families, see also Rawick, ed., *American Slave, Texas,* 4: part 1, 34, 174; *Texas,* 4: part 2, 47, 164, 193; *Texas,* 5: part 3, 85; *Texas,* 5: part 4, 3, 147; *We Are Your Sisters: Black Women in the Nineteenth Century,* ed. Dorothy Sterling (New York: W. W. Norton & Company, 1984), 11.

68. Marianne and Thomas Gaillard to John L. Palmer, 4 June 1844, Marianne Gaillard to "My dear Brother," 28 November 1844, Palmer Family Papers, USC.

Chapter 4. A Little More of This World's Goods: Family, Kinship, and Economics

1. Reuben Grigsby to Lucian P. Grigsby, 13 December 1836, Abner J. Grigsby and G. H. Grigsby to Reuben Grigsby, 15 January 1846, Grigsby Family Papers, VHS; Mary Boykin Williams Harrison Ames, "Childhood Recollections," 4, USC, cited with permission of Martha Daniels.

2. On household size, see Ann Williams Boucher, "Wealthy Planter Families in Nineteenth-Century Alabama" (Ph.D. dissertation, University of Connecticut, 1978), 50; James E. Davis, *Frontier America, 1800–1840: A Comparative Demographic Analysis of the Frontier Process* (Glendale, Calif.: Arthur H. Clark Company, 1977), 69, Table 1 (six persons, census of 1840); Joseph Karl Menn, "The Large Slaveholders of the Deep South, 1860" (Ph.D. disseration, University of Texas, 1964), 197; Barnes F. Lathrop, *Migration into East Texas, 1835–1860: A Study from the United States Census* (Austin: Texas State Historical Association, 1949), 67–69. The average number of slaves owned by these 290 heads of households was forty-six (lower than the average of fifty-three for seaboard planters). Lewis Cecil Gray, *History of the Agriculture of the Southern United States to 1860,* 2 vols. (Baltimore: Waverly Press, 1933), 1:534, finds median slaveholdings of forty-nine slaves in the Alabama Black Belt and between twenty-eight and forty-four slaves for selected Mississippi counties in 1860. Mary E. Stovall shows that nuclear families were the norm in 151 white households of all social classes from three counties (Adams County, Mississippi, Greene and Shelby counties, Tennessee) in the federal census of 1850; see "Kinship, Family Structure and Dynamics among White Families in the Central South, 1850–1880," 3–4, paper presented at the annual meeting of Organization of American Historians, April 1989. Davis, *Frontier America,* implies that white households of all social classes were nuclear across the entire South in the years 1800 to 1840 but does not discuss structure per se. I thank Professor Stovall for permitting me to cite her paper.

3. Gayle Diary, 15 July 1830, Bayne and Gayle Family Papers, UNC.

4. Elvira Crenshaw to Dorothea Winston, 20 August 1849, Winston Family Papers, MS.

5. F. A. Polk to Sarah Polk, 25 March 1839, Gale and Polk Family Papers, UNC; Mary B. G. Rives to Maria O. Rives, 3 June 1853, Roberts Family Papers, VHS; K. McKenzie to Duncan McLaurin, n.d. October 1848, Duncan McLaurin Papers, DU.

6. Thomas T. and Anna Meade to Harriet E. Brown, 18 March 1849, Whitaker

and Meade Family Papers, UNC; S. A. Finley to Caroline G. Hackett, 8 August 1860, Gordon and Hackett Family Papers, UNC; M. E. A. Drake to Gray and Louisa Sills, 14 June 1839, Jelks Sills Letters, DU.

7. S. A. Finley to Caroline L. Gordon, 17 February 1853, S. A. Finley to "Dear sister," 27 November 1858, Gordon and Hackett Family Papers, UNC; Mary B. G. Rives to Maria O. Rives, 3 June 1853, Roberts Family Papers, VHS; M. E. A. Drake to Gray Sills and Louisa Sills, 8 October 1840, Jelks Sills Letters, DU; Adelaide Crain to Caroline Gordon, 7 November 1857, 3 April 1857, Gordon and Hackett Family Papers, UNC.

8. M. C. Hill to Sarah L. Hill, 12 January 1848, Daniel S. Hill Papers, DU. In David E. Whisnant's perceptive study of folk festivals in twentieth-century Appalachia, he remarks that members of traditional societies can eagerly seek change or celebrate it, but I believe that change is much easier to accept when individuals have access to resources; planter women lost their most important resources—their female kinfolk—when their families migrated. See *All That Is Native and Fine: The Politics of Culture in an American Region*, Fred W. Morrison Series in Southern Studies (Chapel Hill: University of North Carolina Press, 1983), 261-62.

9. *Family Letters of the Three Wade Hamptons, 1782-1901*, ed. Charles E. Cauthen, South Caroliniana Sesquicentennial Series, No. 4 (Columbia: University of South Carolina Press, 1953), 42-43, Wade Hampton II to Mary Fisher Hampton, 8 February 1857; Mary Austin Holley, *Texas* (Austin: published by Texas State Historical Association in cooperation with Center for Studies in Texas History, University of Texas at Austin, 1985), 188; Joseph Holt Ingraham, *The South-West. By a Yankee*, 2 vols., March of America Facsimile Series, No. 76 (New York: Harper & Brothers, 1835; repr. Ann Arbor, University Microfilms, 1966), 2:161; Marianne Gaillard to John L. Palmer, 4 June 1844, Palmer Family Papers, USC.

10. John W. DuBose, "A Memoir of Four Families," 2:231, 227, John Witherspoon DuBose Papers, ALA; Herbert Weaver, *Mississippi Farmers, 1850-1860* (Nashville: Vanderbilt University Press, 1945), 80; Paul W. Gates, *The Farmer's Age: Agriculture 1815-1860*, Economic History of the United States, vol. 3 (New York: Holt, Rinehart and Winston, 1960), 146-47; *Agriculture of the United States in 1860, Compiled from the Original Returns of the Eighth Census*, comp. Joseph G. Kennedy (Washington, D.C.: Government Printing Office, 1864), 218, 193. I recalculated Gates's figures for Dallas, Marengo, Montgomery, and Greene counties in Alabama's Black Belt. The difference in size between Charlotte County (468 square miles) and Marengo County (743 square miles) is not sufficient to explain the many large plantations in Marengo; see Joseph Nathan Kane, *The American Counties*, 4th ed. (Metuchen, N.J.: Scarecrow Press, 1983), 80, 202. Robert C. Kenzer, *Kinship and Neighborhood in a Southern Community: Orange County, North Carolina, 1849-1881* (Knoxville: University of Tennessee Press, 1987), 10, estimates that there was one family for every two hundred acres of land in Orange County in 1850. Frederick F. Siegel, *The Roots of Southern Distinctiveness: Tobacco and Society in Danville, Virginia, 1780-1865* (Chapel Hill: University of North Carolina Press, 1987), 78, finds that the average plot size in Pittsylvania and Augusta counties, Virginia, ranged from 150 to 200 acres according to tax lists for 1820, 1840, and 1860. Lewis Gray notes that the average farm (not plantation) in the

seaboard in 1850 was larger (399 acres) than the average farm in the Southwest (273 acres) but excludes Texas from his calculations; see *History of Agriculture,* 1:530.

11. S. A. Townes to G. F. Townes, 14 September 1835, Townes Family Papers, USC; Gayle Diary, 21 June 1831, Bayne and Gayle Family Papers, UNC; H. B. Eggleston to "My Dear Cousin," 13 April 1859, Archer Papers, UT.

12. *Letters from Alabama 1817–1822 by Anne Newport Royall,* biographical introduction and notes by Lucille Griffith, Southern Historical Publications No. 14 (University, Ala.: University of Alabama Press, 1969), Anne Royall to "Matt," 3 May 1821, 217; Ingraham, *South-West* 2:171; "A Journey through the South in 1836: Diary of James D. Davidson," ed. Herbert A. Kellar, *Journal of Southern History* 1 (February–November 1935): 372, 3 December 1836; Holley, *Texas,* 94–100; *Papers of John C. Calhoun,* Volume 15, *1839–1841,* ed. Clyde N. Wilson (Columbia: published by University of South Carolina Press for South Carolina Department of Archives and History and South Carolinana Society, 1983), 342–43, Andrew P. Calhoun to John C. Calhoun, 3 September 1840.

13. R. R. Hackett to R. F. Hackett, 13 April 1849, Gordon and Hackett Family Papers, UNC; *The American Slave: A Composite Autobiography,* ed. George P. Rawick, Contributions in Afro-American and African Studies, No. 11 (Westport, Conn.: Greenwood Publishing Company, 1972), *Texas,* 5: part 4, 106; Journal of Eliza Irion, Book 1, n.d. December 1860, 2, Irion-Neilson Papers, MS. On traveling with male escorts, see Mary E. Meade to Harriet E. Meade, 31 March 1830, Whitaker and Meade Family Papers, UNC; Mary Ann Taylor to Marion Singleton, n.d. January 1834, Singleton Family Papers, LC; [Mrs. Stephen D. Miller], Biographical fragment on Stephen D. Miller, 3, Chesnut-Miller-Manning Papers, SCHS.

14. Mary Ann Whitfield to Gaius Whitfield, 31 March 1835, Gaius Whitfield Papers, ALA.

15. Hardy V. Wooten Diary, 2:41, 23 April 1840, Hardy Vickers Wooten Papers, ALA; M. E. A. Drake to Gray and Louisa Sills, 14 June 1839, Jelks Sills Letters, DU; S. A. Townes to G. F. Townes, 29 September 1836, Townes Family Papers, USC; Gayle Diary, 18 May 1831, 21 June 1831, Bayne and Gayle Family Papers, UNC; Mary Ann S. Black to Sarah Dromgoole, 20 April 1844, Edward Dromgoole Papers, UNC. See also J. B. Downing to "My dear Cousin," 16 February 1857, Archer Papers, UT. Elizabeth Fox-Genovese, *Within the Plantation Household: Black and White Women of the Old South,* Gender and American Culture Series, ed. Linda K. Kerber and Nell Irvin Painter (Chapel Hill: University of North Carolina Press, 1988), 1–28, portrays the Gayle marriage in a more positive light. As I read Sarah Gayle's diary, it suggests that she loved her husband but nonetheless felt frustrated and hurt by his arbitrary decisions and long absences from home.

16. Gayle Diary, 23 June 1830, 10 September 1831, Bayne and Gayle Family Papers, UNC; Caroline Gordon to R. F. Hackett, 16 May 1859, Gordon and Hackett Family Papers, UNC; S. A. Finley to her sister, 9 April 1856, Hamilton Brown Papers, UNC. See also F. A. Polk to Sarah Polk, 25 March 1839, Gale and Polk Family Papers, UNC.

17. Emma Morehead Whitfield, *Whitfield, Bryan, Smith, and Related Families,*

ed. Theodore Marshall Whitfield, 2 vols. (n.p., 1948), 1: 89-93, 298-99. See also Woodruff Narrative, 327, Harris Papers, SCHS.

18. *Memorials of a Southern Planter: By Susan Dabney Smedes*, ed. Fletcher M. Green (New York: Alfred A. Knopf, 1965), 168; R. R. Hackett to R. F. Hackett, 20 November 1850, L. A. Gray to R. F. Hackett, 11 January 1858, Gordon and Hackett Family Papers, UNC; Anna M. Gayle Fry, *Memories of Old Cahaba* (Nashville: Publishing House of M.E. Church, South, 1908), 18, 42-43.

19. Yi-Fu Tuan, "Place: An Experiential Perspective," *Geographical Review* 65 (April 1975): 164; "Martha" to her mother, 1 January 1858, Scarborough Family Papers, DU. On the use of geographic space, see John Brinckerhoff Jackson, *The Necessity for Ruins and Other Topics* (Amherst: University of Massachusetts Press, 1980), 16; Conrad M. Arensburg and Solon T. Kimball, *Culture and Community*, general editor Robert K. Merton (New York: Harcourt, Brace & World, 1965), 3. The stimulating literature by geographers has been strangely neglected by historians, but it offers many insights on the settlement process.

20. Needham Whitfield to Gaius Whitfield, 30 April 1837, Gaius Whitfield Papers, ALA; [Henry Tayloe] to "Dear Brother," 28 January 1837, Tayloe Family Papers, UVA; Larry Schweikart, *Banking in the American South from the Age of Jackson to Reconstruction* (Baton Rouge: Louisiana State University Press, 1987), 60-64, 69, 148-58, 175-82; Peter Temin, *The Jacksonian Economy* (New York: W. W. Norton & Company, 1969), 136-47; Schweikart, "Alabama's Antebellum Banks: New Interpretations, New Evidence," *Alabama Review* 38 (July 1985): 202-21; Gray, *Agriculture*, 2:894-907, 1027. See also Josiah Nott to James H. Hammond, 15 February 1837, Hammond Papers, USC; Duncan McKenzie to Duncan McLaurin, 26 April 1840, Duncan McLaurin Papers, DU. Schweikart points out that private bankers operated in the Southwest after the Panic of 1837, supplying capital after many state banks collapsed, but none of the migrants in this study borrowed from them. See "Private Bankers in the Antebellum South," *Southern Studies* 25 (Summer 1986): 125-34.

21. John L. T. Sneed to William Gaston, n.d. 1843, William Gaston Papers, UNC; John Kerr to Andrew Funkhouser, 23 August 1858, 1 March 1859, Funkhouser Papers, DU.

22. S. A. Townes to G. F. Townes, 8 September 1834, S. A. Townes to J. A. Townes, n.d. August 1836, S. A. Townes to H. H. Townes, 10 September 1839, 28 May 1840, 22 May 1844, 13 May 1845, 27 May 1845, S. A. Townes to G. F. Townes, 15 January 1840, J. A. Townes to W. A. Townes, 4 January 1845, S. A. Townes to W. A., Townes, 4 January 1845, 25 January 1845; H. H. Townes to Rachel Townes, 16 January 1846, S. A. Townes to H. H. Townes, 23 June 1847, H. H. Townes to W. A. Townes, 1 July 1847, H. H. Townes to Rachel Townes, 1 July 1847, H. H. Townes to S. A. Townes, 1 February 1848, Townes Family Papers, USC; "History of Marion, Perry County, by S. A. Townes," *Alabama Historical Quarterly* 14 (no month, 1952): 183, 202-9; Sixth Decennial Census of the United States, Alabama, Perry County, 270. Retinal disease or a mild stroke may have caused Townes's blindness, or this may have been a case of "hysterical blindness" or "conversion blindness," a reaction to the anxiety he was experiencing. See Michael I. Weintraub, *A Clinician's Manual of Hysterical Conversion Reactions* (New York: Interdisciplinary Communications Media, 1978).

23. James B. Whitfield to Needham Whitfield, 30 September, 3 October 1837, Whitfield and Wooten Family Papers, UNC; Joseph B. Anderson to Richard T. Archer, 7 May 1845, E. F. Eggleston to Richard T. Archer, 20 March 1849, Richard T. Archer to Ann Archer, 12 January 1850, Archer Papers, UT; Schweikart, *Banking*, 145–58, 175–82.

24. *Correspondence of James K. Polk*, Volume 4, *1837–1838*, ed. Herbert Weaver, assoc. ed. Wayne Cutler, 6 vols. (Nashville: Vanderbilt University Press, 1977), 4: 279–81, William H. Polk to James K. Polk, 2 December 1837; George Carleton to H. H. Worthington, 14 January 1847, Edward Dromgoole Family Papers, UNC; Abner J. Grigsby to Reuben Grigsby, 6 April 1848, Grigsby Family Papers, VHS; *The Lides Go South . . . And West: The Records of a Planter Migration in 1835*, ed. Fletcher M. Green (Columbia: University of South Carolina Press, 1952), Margaret Lide to James Lide, 21 May 1854, 43.

25. On hiring slaves, see James P. Cocke to Richard Archer, 4 November 1839, William Eggleston to Richard Archer, 29 October 1842, Stephen H. Eggleston to Richard Archer, 13 April 1850, Archer Papers, UT; Samuel T. Nicholson to Blake Nicholson, 14 August 1854, Nicholson Papers, MS. On inheritance, see William B. Inge to John Bullock, 8 January 1841, John Bullock and Charles E. Hamilton Papers, UNC; John McCreary to "Dear Brother," 18 March 1858, Gaston-Crawford Papers, USC.

26. Willis Lea to William Lea, 1 February 1853, Lea Family Papers, UNC; John S. Burwell to Lewis Burwell, 11 September 1845, Lewis Burwell Papers, DU; Duncan McKenzie to Duncan McLaurin, 3 April 1842, Duncan McLaurin Papers, DU.

27. O. G. Murrell to John Murrell, 16 July 1849, O. G. and Emily Murrell to John Murrell, 22 July 1849, Murrell Family Papers, UVA; D. G. Rencher to William Merritt, 13 December 1849, Abraham Rencher Papers, UNC.

28. Adelaide Crain to Caroline Gordon, 28 June 1856, S. A. Finley to Caroline G. Hackett, 8 August 1860, S. A. Finley to Caroline G. Hackett, 5 December 1860, Gordon and Hackett Family Papers, UNC; Gayle Diary, 14 March 1828, Bayne and Gayle Family Papers, UNC. See also O. G. Murrell and Emily Murrell to John Murrell, 22 July 1849, Murrell Family Papers, UVA. Robert V. Wells, *Revolutions in Americans' Lives: A Demographic Perspective on the History of Americans, Their Families, and Their Society*, Contributions in Family Studies, No. 6 (Westport, Conn.: Greenwood Press, 1982), 93, notes that couples in frontier areas began limiting the size of their families later in the nineteenth century than couples on the Northern and Southern seaboard.

29. M. E. A. Drake to Gray and Louisa Sills, 14 June 1839, Jelks Sills Letters, DU; Gayle Diary, 4 September 1829, Bayne and Gayle Family Papers, UNC; M. E. A. Drake to Gray Sills and Louisa Sills, 8 October 1840, Jelks Sills Letters, DU. Women on other frontiers wanted female company during childbirth; see Stratton, *Pioneer Women*, 86; Julie Roy Jeffrey, *Frontier Women: The Trans-Mississippi West, 1840–1880*, American Century Series (New York: Hill and Wang, 1979), 69.

30. Martha Pickens to Anna Lenoir Jones, 29 April 1820, "L. C. N." [Laura Norwood] to Julia P. Howe, 12 March 1839, biographical fragments on Julia

Pickens Howe, n.d., Israel Pickens Papers, ALA; Thomas Felix Hickerson, *Echoes of Happy Valley: Letters and Diaries, Family Life in the South, Civil War History* (Chapel Hill: by the author, 1962), 14, idem, *Happy Valley: History and Genealogy* (Chapel Hill: by the author, 1940), 106.

31. Ann Archer to "My dear Child," 3 April 1854, Archer Papers, UT; Ann Finley to Caroline Gordon, 26 December 1853, Adelaide Crain to Caroline Gordon, 3 April 1857, Gordon and Hackett Family Papers, UNC.

32. All percentages are rounded. The average number of slaves owned by these fathers was ninety-eight, and the median number was sixty-two. One planter's son, Charles Tayloe, left the South and pursued a diplomatic career, although he owned slaves in Virginia; see Table 2. Some scholars find that migrants did better economically than nonmigrants. Lee Soltow, *Men and Wealth in the United States, 1850-1870* (New Haven: Yale University Press, 1975), 194 n.24, estimates that the average wealth for whites in the Southwest (including Kentucky and Tennessee) was "about 5-10 percent" greater than that of whites in the Southeast in 1860 and 1870. Econometric historians argue that the growth of per capita income in the Southwest was higher in the years 1840 to 1860 than in the seaboard. For a summary of this literature, see Robert E. Gallman, "Slavery and Southern Economic Growth," *Southern Economic Journal* 45 (April 1979): 1007-22. Peter R. Knights, *The Plain People of Boston, 1830-1860: A Study in City Growth* (New York: Oxford University Press, 1971), 115-18, compares a sample of 31 men who migrated from Boston and 142 persisters who stayed behind in 1850 and 1860; 20 percent of the persisters increased their holdings in real estate and personal property, while 40 percent of the migrants increased their wealthholdings.

Yet other historians stress the disadvantages of migration. Hal S. Barron, *Those Who Stayed Behind: Rural Society in Nineteenth-century New England* (Cambridge: Cambridge University Press, 1984), 97-111, suggests that farmers in Chelsea, Vermont, prospered partly because of aid from relatives, but he does not compare their holdings with those of migrants from Chelsea. Jane Turner Censer compares slaveholdings of North Carolina planters who owned at least seventy slaves in 1830 with those of their adult offspring in 1850 and 1860. She finds that most children (two-thirds in 1850 and three-fourths in 1860) became planters but only about one-fourth owned seventy or more slaves in 1860. See *North Carolina Planters and Their Children, 1800-1860* (Baton Rouge: Louisiana State University Press, 1984), 123-25. Stephan Thernstrom, *Poverty and Progress: Social Mobility in a Nineteenth-Century City* (New York: Atheneum Press, 1977), 86-114, believes that white working-class men who left Newburyport, Massachusetts, between 1850 and 1880 probably did not experience the upward occupational mobility of those who remained behind.

33. Will of James Calhoun, probated 21 February 1843, Abbeville County Estate Papers, Box 23, Package 526, SCA. The seventh father, William Lea, died in 1873. See Table 2.

34. Andrew L. Pickens to C. S. Howe, 20 July 1840, James Pickens to Samuel Pickens, 17 February 1846, Israel Pickens Papers, ALA; Map of Greene County, Alabama, by V. Gayle Snedecor, 1858, esp. precinct 6, ALA; Marilyn Davis Hahn, *Old Cahaba Land Office Records & Military Warrants 1817-1863* (Mobile: Old

South Printing & Publishing Company, 1981); Fry, *Memories of Cahaba,* 66; *Correspondence of James K. Polk, Volume 1, 1817–1832,* ed. Herbert Weaver, assoc. ed. Paul H. Bergeron, 6 vols. (Nashville: Vanderbilt University Press, 1969), 1:152, Marshall T. Polk to James K. Polk, 20 February 1828, 324, William Polk to James K. Polk, n.d. August 1830; 2:221–22, James K. Polk to William Polk, 5 January 1834; 3:332–33, William J. Polk to James K. Polk, 15 October 1835; Gaius Whitfield to Allen W. Wooten, 9 August 1843, 12 November 1850, Needham Whitfield to Allen W. Wooten, 5 March 1850, 18 June 1856, Whitfield and Wooten Family Papers, UNC. Morton Rothstein, "The Natchez Nabobs: Kinship and Friendship in an Economic Elite," in *Toward a New View of America: Essays in Honor of Arthur C. Cole,* ed. Hans L. Trefousse (New York: Burt Franklin and Company, 1977), 97–112, argues that familial assistance was essential to the success of four extremely wealthy Natchez planters and businessmen in the Duncan, Marshall, Mercer, and Minor families. (None of these men is in Table 2.)

35. Seventh Decennial Census of the United States, Free Schedule, Kentucky, Barren County, 485, Slave Schedule, Kentucky, Barren County, n.p.; Marion N. Carr to George Carr, 8 March 1852, Carr Papers, UVA.

36. Randolph B. Campbell and Richard G. Lowe, *Wealth and Power in Antebellum Texas* (College Station: Texas A & M University Press, 1977), 54–56; William L. Barney, "Patterns of Crisis: Alabama White Families and Social Change, 1850–1870," *Sociology and Social Research* 63 (1979): 527–28, 530; Barney, *The Secessionist Impulse: Alabama and Mississippi in 1860* (Princeton, N.J.: Princeton University Press, 1974), 3–5; Gavin Wright, *The Political Economy of the Cotton South: Households, Markets, and Wealth in the Nineteenth Century* (New York: W. W. Norton & Company, 1978), 29–37 (percentages on 34). Cf. James Oakes, *The Ruling Race: A History of American Slaveholders* (New York: Alfred A. Knopf, 1982), 37–41, who stresses opportunities for whites to own slaves and the "direct material interest" that many whites had in perpetuating slavery rather than the concentration of wealth in Southern society.

37. Marianne Gaillard to John L. Palmer, 8 August 1844, Palmer Family Papers, USC.

38. Joseph H. Parks, *General Leonidas Polk C.S.A., the Fighting Bishop* (Baton Rouge: Louisiana State University Press, 1962), 80; Mary Bray Wheeler and Genon Hickerson Neblett, *Chosen Exile: The Life and Times of Septima Sexta Middleton Rutledge, American Cultural Pioneer* (Gadsden, Ala.: Rutledge Hill Press, 1980), 107; Martha Gamble to Jane G. B. Crawford, 9 March 1860, Gaston-Crawford Papers, USC.

39. Gayle Diary, 20 October 1827, Bayne and Gayle Family Papers, UNC; M. E. A. Drake to Gray and Louisa Sills, 14 June 1839, Jelks Sills Letters, DU; "Martha" to her mother, 1 January 1858, Scarborough Family Papers, DU; Hickerson, *Happy Valley,* 111.

40. Hickerson, *Happy Valley;* "Martha" to her mother, 1 January 1858, Scarborough Family Papers, DU; M. E. A. Drake to Gray Sills and Louisa Sills, 14 June 1839, Jelks Sills Letters, DU.

41. S. A. Townes to G. F. Townes, 8 September 1834, S. A. and Joanna Townes to J. A. Townes, 18 September 1834, S. A. Townes to Rachel Townes, 2 November

1834, Joanna Townes to G. F. Townes, 11 February 1838, H. H. Townes to G. F. Townes, 24 August 1847, H. H. Townes to W. A. Townes, 18 December 1847, Townes Family Papers, USC. The family did return in the late 1840s.

42. M. E. Perkins to Caroline Gordon, 1 May 1857, Gordon and Hackett Family Papers, UNC.

43. Hickerson, *Happy Valley*, 112; M. L. Finley to Caroline Gordon, 26 August 1853, Gordon and Hackett Family Papers, UNC; *Memoirs of Maverick*, 14, 63.

44. Sarah Kelsea to "My Beloved Friend," 7 April 1856, Norton Papers, USC; Adelaide Crain to Caroline Gordon, 27 December 1854, Gordon and Hackett Family Papers, UNC.

45. Duncan McKenzie to Duncan McLaurin, 26 August 1844, Duncan McLaurin Papers, DU; "History of Marion," 183.

46. Henry A. Tayloe to Benjamin O. Tayloe, n.d. 1836, Tayloe Family Papers, UVA; Columbus Morrison to James Morrison, 29 January 1837, Bondurant-Morrison Papers, UVA; L. K. Person to Thomas A. Person, 10 January 1839, Person Letters and Papers, DU.

47. William D. Gale to Ann M. Gale, 8 April 1844, Gale and Polk Family Papers, UNC.

Chapter 5. To Live Like Fighting Cocks: Independence, Sex Roles, and Slavery

1. Migration did not reduce differences between the Southern seaboard and the frontier, as Frederick Jackson Turner suggested, but heightened them. See "The Significance of the Frontier in American History," in *The Turner Thesis Concerning the Role of the Frontier in American History*, 3rd ed., ed. George Rogers Taylor (Lexington, Mass.: D. C. Heath, 1972).

2. Letter to the Editor, *Hinds County (Miss.) Gazette*, 23 April 1856; Gayle Diary, 18 May 1831, n.d. November 1828, Bayne and Gayle Family Papers, UNC; Caroline L. Gordon to her brother, 20 May 1853, James Gordon Hackett Papers, DU; Ann Finley to Caroline Gordon Hackett, 8 August 1860, Gordon and Hackett Family Papers, UNC. See also Annie Archer to Ann Archer, 31 July 1858, Archer Papers, UT; Mary Bray Wheeler and Genon Hickerson Neblett, *Chosen Exile: The Life and Times of Septima Sexta Middleton Rutledge, American Cultural Pioneer* (Gadsden, Ala.: Rutledge Hill Press, 1980), 73.

3. Frederick Law Olmsted, *The Cotton Kingdom: A Traveller's Observations on Cotton and Slavery in the American Slave States*, ed. Arthur M. Schlesinger, Sr., introduction by Lawrence N. Powell (New York: Modern Library, 1984), 414, 426; *From Virginia to Texas, 1835: Diary of William F. Gray Giving Details of His Journey to Texas and Return in 1835–1836 and Second Journey to Texas in 1837*, preface by A. C. Gray (Houston: Gray, Dillaye & Co., 1909; repr. Houston: Fletcher Young Publishing Co., 1965), 40; S. H. Aby to his parents, 15 February 1840, Aby Family Papers, MS. See also M. E. A. Drake to Gray and Louisa Sills, 14 June 1839, Jelks Sills Letters, DU; B. B. Ball to Lucinda Robinson, 24 November 1850, Ball Family Papers, VSA; [Ann?] Finley to Caroline Gordon, 26 December 1853, Gordon and Hackett Family Papers, UNC.

4. "Moral Education," *DeBow's Review* 18 (March 1855): 432. For other editorials, see "How to Ruin a Son," *Gainesville (Ala.) Independent*, 26 Feb. 1859; "A Few Words to Parents," *Yazoo City (Miss.) Weekly Whig*, 21 Jan. 1853; editorial, *Yazoo City (Miss.) Weekly Whig*, 20 Nov. 1858; editorial, *Houston Telegraph and Texas Register*, 14 July 1841, quoted in William Ransom Hogan, *The Texas Republic: A Social and Economic History* (Norman: University of Oklahoma, 1946), 143. DeBow's sons were small boys in the 1850s. See Otis Clark Skipper, *J. D. B. DeBow: Magazinist of the Old South* (Athens: University of Georgia Press, 1958), 108, 110, 172.

5. Thomas Aby to his parents, 7 March 1847, S. H. Aby to his parents, 19 October 1847, Aby Family Papers, MS; S. A. Townes to Rachel Townes, 9 April 1843, Townes Family Papers, USC; *Papers of John C. Calhoun*, Volume 15, *1839–1841*, ed. Clyde W. Wilson (Columbia: published by University of South Carolina Press for South Carolina Department of Archives and History and South Caroliniana Society, 1983), 340, Andrew Pickens Calhoun to John C. Calhoun, 28 August 1840.

6. Joanna Townes to W. A. Townes, n.d. November 1844, S. A. Townes to G. F. Townes, 28 December 1841, S. A. Townes to Rachel Townes, 22 June 1843, Townes Family Papers, USC; Gayle Diary, 9 March 1828, 22 April 1828, Bayne and Gayle Family Papers, UNC.

7. Gayle Diary, 20 July 1828, Bayne and Gayle Family Papers, UNC; Ray Mathis, *John Horry Dent: South Carolina Aristocrat on the Alabama Frontier* (University: published under the sponsorship of the Historic Chattahoochee Commission by University of Alabama Press, 1979), 167, 41; *Diary of William Gray*, 227.

8. Joanna Townes to G. F. Townes, 11 February 1838, S. A. Townes to Rachel Townes, 22 June 1843, S. A. Townes to "My Dear Brother" and Rachel Townes, 17 September 1843, S. A. Townes to H. H. Townes, 17 August 1844, S. A. Townes to Rachel Townes, 27 May 1845, S. A. Townes to Rachel Townes, n.d., Townes Family Papers, USC. John Blassingame either died or disappeared in the 1840s. He may have gone to Mississippi in 1845 after he stabbed a man in Marion; see S. A. Townes to H. H. Townes, 13 May 1845, Townes Family Papers, USC.

9. S. A. Townes to John and G. F. Townes, 1 September 1834, S. A. Townes to W. A. Townes, 25 June 1841, Townes Family Papers, USC. The sport of cockfighting had been popular in the South since the colonial era; scholars disagree on its social and cultural meaning. Rhys Isaac, *The Transformation of Virginia, 1740–1790* (Chapel Hill: published for Institute of Early American History and Culture, Williamsburg, Virginia, by University of North Carolina Press, 1982), 101–4, argues that gatherings at cockfights confirmed social distinctions within the white population, while Elliott J. Gorn, "'Gouge and Bite, Pull Hair and Scratch': The Social Significance of Fighting in the Southern Backcountry," *American Historical Review* 90 (February 1985): 18–43, portrays cockfighting as a distinctly nonelite sport practiced by yeomen and backwoodsmen. Samuel Townes did not mention attending cockfights, but his comments lend support to Bertram Wyatt-Brown's view that the sport was popular among all social classes and gave men a sense of power; see *Southern Honor: Ethics and Behavior in the Old South* (New York: Oxford University Press, 1982), 340–44. For a colorful description of cockfighting

in the South today, see Pete Daniel, *Standing at the Crossroads: Southern Life Since 1900*, American Century Series, consulting editor Eric Foner (New York: Hill and Wang, 1986), 189–93.

10. W. J. Rorabaugh, *The Alcoholic Republic, an American Tradition* (New York: Oxford University Press, 1979), 14; *The American Slave: A Composite Autobiography*, ed. George P. Rawick, Contributions in Afro-American and African Studies, No. 11 (Westport, Conn.: Greenwood Publishing Company, 1972), *Alabama*, 6: 312–13; "The Importance of Self-Government," *Natchez Southwestern Journal*, 30 December 1837. Simon Phillips condoned the heavy drinking among planter men. See also *Diary of William Gray*, 29; *Travels in the Old South: Selected from Periodicals of the Times*, ed. Eugene L. Schwab with the collaboration of Jacquelin Bull, 2 vols. (Lexington: University of Kentucky Press, 1973), 2:531.

11. Fenton Noland to Callendar Noland, 2 February 1853, Noland Family Papers, UVA; D. G. Rencher to William Merritt, 29 January 1852, Abraham Rencher Papers, UNC; *Correspondence of James K. Polk:* Volume 1, *1817–1832*, ed. Herbert Weaver, assoc. ed. Paul H. Bergeron (Nashville: Vanderbilt University Press, 1969): xxxviii, 123–24, Jane Polk to James K. and Sarah Polk, 5 January 1828; S. A. Townes to G. F. Townes, 10 September 1841, H. H. Townes to Rachel Townes, 30 November 1843, Townes Family Papers, USC; S. A. Finley to her sister, 9 April 1856, Hamilton Brown Papers, UNC. For a summary of contemporary controversies about the causes of alcoholism, see George E. Vaillant, *The Natural History of Alcoholism* (Cambridge, Mass.: Harvard University Press, 1983). Isaac, *Transformation*, 95, 114, notes that men in eighteenth-century Virginia proved their "manly prowess" in public places such as taverns and racetracks.

12. S. A. Townes to G. F. Townes, 22 June 1834, Townes Family Papers, USC; George P. Tayloe to Benjamin O. Tayloe, 2 April 1840, Tayloe Family Papers, UVA; John J. Allen, Sr., to John J. Allen, Jr., 16 July 1855, Allen Family Papers, UVA. On horse racing, see also Willis Lea to William Lea, 20 May 1844, Lea Family Papers, UNC; Joseph Holt Ingraham, *The South-West. By a Yankee*, 2 vols., March of America Facsimile Series, No. 76 (New York: Harper & Brothers, 1835; repr. Ann Arbor: University Microfilms, 1966), 2:19–21, 59, 219–20; Edwin Adams Davis, "Introduction," 57–60, in *Plantation Life in the Florida Parishes of Louisiana, 1836–1846, as Reflected in the Diary of Bennet H. Barrow* (New York: AMS Press, 1967). On gambling, see also *Diary of William Gray*, 62; James R. Creecy, *Scenes in the South, and Other Miscellaneous Pieces* (Washington, D.C.: Thomas McGill, 1860), 109–11; "A Journey through the South in 1836: Diary of James D. Davidson," ed. Herbert A. Kellar *Journal of Southern History* 1 (February-November 1935): 356, 5 November 1836; Virginia Park Matthias, "Natchez-Under-the-Hill as It Developed under the Influence of the Mississippi River and the Natchez Trace," *Journal of Mississippi History* 8 (October 1945): 201–21.

13. Anna M. Gayle Fry, *Memories of Old Cahaba* (Nashville, Tenn.: Publishing House of M. E. Church, South, 1908), 25; Johan Huizinga, *Homo Ludens: A Study of the Play Element in Culture* (New York: Harper & Row, 1970), 19–22, 70–73. Wyatt-Brown, *Southern Honor*, 349–50, suggests that compulsive gamblers in the Old South, like their counterparts today, may have been struggling to resolve psychological conflicts with their fathers and bolster their own vulnerable self-

esteem, although he notes that contemporary psychologists do not agree on the origins of compulsive gambling. Wyatt-Brown's path-breaking book is based in large part on families who lived in the Southern interior or the old Southwest. He explores several attributes of the male sex role I discuss but does not tie them to changes wrought by migration.

14. Ingraham, *South-West,* 2:166–68. On duels, see also John Guion to George W. Guion, 30 March 1824, Guion Papers, UNC; Harriet Martineau, *Society in America,* 3 vols. (London: Saunders and Otley, 1837; repr. New York: AMS Press, 1966), 1:308; *Texas in 1837: An Anonymous, Contemporary Narrative,* ed. Andrew Forest Muir (Austin: University of Texas Press, 1958), 38–41, 160; "The Duelist's Doom," *Hinds County (Miss.) Gazette,* 4 October 1854; Fry, *Memories of Cahaba,* 31–32; W. Stuart Harris, "Rowdyism, Public Drunkenness, and Bloody Encounters in Early Perry County," *Alabama Review: A Quarterly Journal of Alabama History* 33 (January 1980): 17. Cf. Steven M. Stowe, *Intimacy and Power in the Old South: Ritual in the Lives of Planters,* New Studies in American Intellectual and Cultural History, Thomas Bender, consulting editor (Baltimore: Johns Hopkins University Press, 1987), 5–49, and Kenneth S. Greenberg, *Masters and Statesmen: The Political Culture of American Slavery,* New Studies in American Intellectual and Cultural History, Thomas Bender, consulting editor (Baltimore: Johns Hopkins University Press, 1985), 23–41, who both depict duels as expressions of the elite social class of planters; only gentlemen could engage in duels. By contrast, Wyatt-Brown, *Southern Honor,* 350–61, downplays elitism and portrays duels as expressions of honor and masculinity. Dickson D. Bruce, Jr., *Violence and Culture in the Antebellum South* (Austin: University of Texas Press, 1979), 1–43, suggests that duels provided ways for Southerners to control and channel dangerous passions. On the volatility of dense populations of strangers in public environments, see Drury R. Sherrod and Sheldon Cohen, "Density, Personal Control, and Design," in *Humanscape: Environments for People,* ed. Stephen Kaplan and Rachel Kaplan (Ann Arbor: Ulrich's Books, 1982), 331–38.

15. Fred Darkis, Jr., "Alexander Keith McClung (1811–1855)," *Journal of Mississippi History* 60 (November 1978): 290–96; Virginia Clay-Clopton, *A Belle of the Fifties: Memoirs of Mrs. Clay, of Alabama, Covering Social and Political Life in Washington and the South, 1853–1866, put into narrative form by Ada Sterling* (New York: Doubleday, Page & Company, 1905), 15–16; Wyatt-Brown, *Southern Honor,* 358–60; *Correspondence of James K. Polk,* Volume 5, *1839–1841,* 6 vols., ed. Wayne Cutler, assoc. eds. Earl J. Smith and Carese J. Parker (Nashville: Vanderbilt University Press, 1979), 5:3–6, William H. Polk to James K. Polk, 2 January 1839.

16. Martineau, *Society in America,* 1:285, 308; S. A. Townes to Rachel Townes, 22 June 1843, Townes Family Papers, USC; *Humor of the Old Southwest,* ed. Hennig Cohen and William B. Dillingham, 2nd ed. (Athens: University of Georgia Press, 1975). Bruce, *Violence and Culture,* 225–32, portrays Southwestern humorists as critics of the violence on the frontier, which they saw as a place where the worst human passions were given free rein. On the violence of planters, see also Timothy Flint, *Recollections of the Last Ten Years, Passed in Occasional Residences and Journeyings in the Valley of the Mississippi,* introduction by James D. Norris, The

American Scene: Comments and Commentators, general ed. Wallace D. Farnham (New York: Da Capo Press, 1968), 337; Creecy, *Scenes in the South*, 67.

17. John G. Guion to George S. Guion, 31 December 1827, Guion Papers, UNC; W. W. Harrison to Benjamin Yancey, 3 December 1837, Benjamin Cudworth Yancey Papers, UNC; Rawick, ed., *American Slave, Texas*, 5: part 3, 46, *Texas*, 4: part 2, 88. See also "An Unfaithful Husband," *Voice of Sumter* (*Livingston, Alabama*), 31 May 1836.

18. Duncan W. McKenzie to John McLaurin, 29 March 1838, Duncan McLaurin Papers, DU; Catherine Clinton, *The Plantation Mistress: Woman's World in the Old South* (New York: Pantheon Books, 1982), 211–12; Gayle Diary, 23 January 1828, 7 February 1828, Bayne and Gayle Family Papers, UNC.

19. Testimony of William Maxwell, 26 September 1850, Case of William J. Wilson, Records of the Montgomery Circuit, Methodist Episcopal Church, Archives of the State of Texas. See also Wyatt-Brown, *Southern Honor*, 311, 320–24. Cf. Joel Williamson, *New People: Miscegenation and Mulattoes in the United States* (New York: Free Press, 1980), 57–59, who argues that miscegenation was uncommon in the Southwest because slaves were physically isolated from whites and sex ratios among whites were roughly equal.

20. "Memoir of Harriet M. P. P. Ames," 6–9, Potter Papers, DU; Duncan C. Calhoun to Duncan McLaurin, 15 May 1835, Duncan McLaurin Papers, DU.

21. *Dictionary of American Biography*, ed. Allen Johnson, 20 vols. (New York: Charles Scribner's Sons, published under the auspices of American Council of Learned Societies, 1928-annually), 1:338–39, 20:35; Pat Ireland Nixon, *The Medical Story of Early Texas, 1528–1853*, foreword by Chauncey D. Leake (Lancaster, Pa.: Lancaster Press, 1946), 323–27; *A Comprehensive History of Texas, 1685–1897*, ed. Dudley G. Wooten, 2 vols. (Dallas: William G. Scarff, 1898), 1:189–91; James E. Winston, "Virginia and the Independence of Texas," *Southwestern Historical Quarterly* 16 (January 1913): 281; W. Randolph, "Genealogy of Archer Family," 50, VSA; Hogan, *Texas Republic*, 269; *Diary of William Gray*, 64; Branch T. Archer to Richard T. Archer, 22 October 1854, 21 October 1853, Archer Papers, UT; Seventh Annual Decennial Census of the United States, Texas, Brazoria County, Free Schedule, 387, Slave Schedule; Ann Archer to Edward Archer, 25 March 1855, Archer Papers, UT; *The 1860 Census of Brazoria County, Texas*, trans. Nanetta Key Burkholder (Brazosport: Brazosport Genealogical Society, 1978), 20.

22. J. W. Calvert to John L. Trone, 27 July 1851, Trone Paper, UNC; Cloud Memoirs, 12, 14, Isaac Newton Cloud Memoirs, UVA.

23. H. H. Townes to Rachel Townes, 23 August 1845, Townes Family Papers, USC. See also John B. Dabney Manuscript, 165–69, UVA; William P. Duval to James Madison, 14 November 1826, Ambler and Barbour Family Papers, UVA.

24. Gayle Diary, 3 October 1830, 20 October 1827, 6 November 1831, 12 November 1831, 29 April 1828, 2 December 1827, n.d. September 1827, Bayne and Gayle Family Papers, UNC. For some of the edited passages, see 1 January 1828, 5 March 1828, 13 April 1828, and 20 March 1829.

25. Sandra Moncrief, "The Mississippi Married Women's Propery Act of 1839," *Journal of Mississippi History* 17 (May 1985): 110–25; James W. Ely, Jr., and David

J. Bodenhamer, "Regionalism and the Legal History of the South," in *Ambivalent Legacy: A Legal History of the South*, ed. Bodenhamer and Ely (Jackson: University Press of Mississippi, 1984), 12; Suzanne D. Lebsock, "Radical Reconstruction and the Property Rights of Southern Women," *Journal of Southern History* 42 (May 1977): 195–216; Norma Basch, *In the Eyes of the Law: Women, Marriage, and Property in Nineteenth-Century New York* (Ithaca: Cornell University Press, 1982), 27; Albie Sachs and Joan Hoff Wilson, *Sexism and the Law: A Study of Male Beliefs and Legal Bias in Britain and the United States*, Law in Society Series, ed. C. M. Campbell and P. N. P. Wiles (Oxford: Martin Robertson & Company Ltd., 1978), 77–79. Historians disagree on whether white women's status decreased on other American frontiers; for a convincing discussion of the topic, see John Mack Faragher, *Women and Men on the Overland Trail* (New Haven: Yale University Press, 1979), 62–65. Ann Williams Boucher, "Wealthy Planter Families in Nineteenth-Century Alabama" (Ph.D. dissertation, University of Connecticut, 1978), 127–35, finds that property and divorce laws in Alabama were more lenient toward women than similar laws in Virginia, North Carolina, and South Carolina.

26. Mathis, *John Horry Dent*, 167; Gayle Diary, 21 June 1829 (quotation), 2 December 1827, 4 March 1830, Bayne and Gayle Family Papers, UNC; John W. DuBose, "A Memoir of Four Families," 1:205. Mr. DuBose thought that this incident was amusing.

27. Dabney Manuscript, 168–69, UVA.

28. Ellen Hard Townes, "Genealogy of the Townes Family," 1–2; Lillian Adele Kibler, *Benjamin F. Perry, South Carolina Unionist* (Durham: Duke University Press, 1946), 119; Grantee Index to Conveyances, T-Z, 1787-1913, Greenville County, 1829, Book Q, 120, SCA; S. S. Crittenden, *The Greenville Century Book* (Greenville, S.C.: Greenville News, 1903), 16–19; Sixth Decennial Census of the United States, South Carolina, Perry County, 273; Will of William E. Blassingame, written 13 March 1841, Estate Records of Perry County, Alabama, ALA; S. A. Townes to "My Dear Brother," 22 November 1826, S. A. Townes to G. F. Townes, 18 July 1833, 22 June 1834, 7 October 1834, Samuel and Joanna Townes to G. F. Townes, 10 March 1835, S. A. Townes to G. F. Townes, 10 December 1835, S. A. Townes to J. A. Townes, n.d. August 1836, S. A. Townes to H. H. Townes, 10 September 1839, 9 April 1840, S. A. Townes to G. F. Townes, 28 December 1841, 1 March 1842, S. A. Townes to Rachel Townes, 22 June 1843, Townes Family Papers, USC.

29. *Letters from Alabama by Anne Newport Royall*, biographical introduction and notes by Lucille Griffith, Southern Historical Publications, No. 14 (University: University of Alabama Press, 1969), Anne Royall to "Matt," 3 May 1821, 216–18; Gayle Diary, 15 July 1829, Bayne and Gayle Family Papers, UNC. On women visiting, see Gayle Diary, 3 October 1830, Bayne and Gayle Family Papers, UNC; E. A. Benson to N. F. Norton, 25 June 1855, Norton Papers, USC; Richard Wooten to Gaius Whitfield, 23 July 1834, Gaius Whitfield Papers, ALA; Wyatt-Brown, *Southern Honor*, 276.

30. "Memoir of Ames," 1–10, Potter Papers, DU. See also Dabney Manuscript, 166, UVA.

31. Reginald Horsman, *Josiah Nott of Mobile: Southerner, Physician, and Racial Theorist*, Southern Biography Series, ed. William J. Cooper, Jr. (Baton

Rouge: Louisiana State University Press, 1987); *Selected Writings of Henry Hughes: Antebellum Southerner, Slavocrat, Sociologist*, ed. Stanford M. Lyman (Jackson: University Press of Mississippi, 1985), 73–185; *The Ideology of Slavery: Proslavery Thought in the Antebellum South, 1830–1860*, ed. Drew Gilpin Faust (Baton Rouge: Louisiana State University Press, 1981), 14–16, 206–38; Michael Johnson, "Planters and Patriarchy: Charleston, 1800–1860," *Journal of Southern History* 46 (February 1980): 45–46; John McCardell, *The Idea of a Southern Nation: Southern Nationalists and Southern Nationalism, 1830–1860* (New York: W. W. Norton & Company, 1979), 70–84; William Stanton, *The Leopard's Spots: Scientific Attitudes toward Race in America, 1815–1859* (Chicago: University of Chicago Press, 1972), 80–81, 158–59; George M. Fredrickson, *The Black Image in the White Mind: The Debate on Afro-American Character and Destiny, 1817–1914* (New York: Harper & Row, 1971), 75–76, 78–82; William Sumner Jenkins, *Pro-Slavery Thought in the Old South* (Chapel Hill: University of North Carolina Press, 1935), 209–10. In his excellent biography, Horsman emphasizes Nott's transatlantic reputation as a writer and the acceptance of his ideas throughout the United States rather than his Southwestern associations; I wish to respectfully disagree with his view that Nott's "sense of familial obligation was strong" (65). Faust accents Nott's hostility toward clergymen and his personal belligerence; she disagrees with Fredrickson's argument that there were regional differences among proslavery writers (17 n. 39). William Harper, a proslavery writer, lived in Missouri for five years but spent most of his life in South Carolina; see Faust, ed., *Ideology of Slavery*, 78. Henry Hughes was born in Mississippi but was not himself a migrant; see Lyman, ed., *Henry Hughes*, 3–19.

32. Richard Archer to Ann Archer, 18 January 1850, Archer Papers, UT; Fredrika Bremer, *The Homes of the New World: Impressions of America*, 2 vols. (New York: Harper & Brothers, 1853), 2:188–93. Thomas Brown, "Lineage," 63, Ambler-Brown Papers, DU.

33. William H. Tayloe to Benjamin O. Tayloe, 23 January 1845, Tayloe Family Papers UVA; H. H. Townes to G. F. Townes, 14 November 1835, Townes Family Papers, USC. See also Philip Minor to John Minor, 13 June 1827, Minor Papers, UT; John C. Inscoe, *Mountain Masters, Slavery, and the Sectional Crisis in Western North Carolina* (Knoxville: University of Tennessee Press, 1989), 90. William Tayloe must have overcome some of these objections in later years because he sent some of his slaves to Alabama in 1850 and 1860; see Table 2.

34. S. A. Townes to "Dear Brother," 7 October 1829, S. A. Townes and Joanna Townes to J. A. Townes, 18 September 1834, G. F. Townes to J. A. Townes, 3 April 1834, S. A. Townes to J. A. Townes, 1 October 1834, S. A. Townes to G. F. Townes, 24 April 1835, Townes Family Papers, USC. The Townes family still owned Marcellena in 1845; see J. A. Townes to "Dear Brother," 15 February 1845, Townes Family Papers, USC.

35. Norman R. Yetman, *Life Under the 'Peculiar Institution': Selections from the Slave Narrative Collection* (Huntington, N.Y.: Robert E. Krieger, 1976), 174; Rawick, ed., *American Slave, South Carolina*, 3: part 3, 219; *Texas*, 5: part 3, 45; *Texas*, 4: part 2, 77. See also John W. Murrell to John D. Murrell, 31 December 1848, Murrell Family Papers, UVA; *Diary of William Gray*, 57; Rawick, ed.,

American Slave, Texas, 4: part 2, 88; *Texas,* 5: part 3, 133–35, part 4, 174; *Diary of Bennet Barrow,* 85, 109, 160, 239, Olmsted, *Cotton Kingdom,* 452. Edward L. Ayers, *Vengeance and Justice: Crime and Punishment in the Nineteenth-Century American South* (New York: Oxford University Press, 1984), 134, notes that the law codes of the Old Southwest allowed for more "procedural fairness" for slaves in trouble with the law, but these legal changes probably had little impact on the lives of slaves because, as Ayers remarks, each plantation had a law of its own. Cf. Eugene D. Genovese, *Roll, Jordan, Roll: The World the Slaves Made* (New York: Pantheon Books, 1974), 54, who argues that differences within the South dwindled by the late antebellum era, and John Hebron Moore, *The Emergence of the Cotton Kingdom in the Old Southwest: Mississippi, 1770–1860* (Baton Rouge: Louisiana State University Press, 1988), 89–90, who argues that the living conditions of slaves improved after the 1840s as the frontier period ended.

36. S. A. Townes to Rachel Townes, 9 April 1843, Townes Family Papers, USC; Mathis, *John Horry Dent,* 167; Solomon Northup, *Narrative of Solomon Northup, a Citizen of New York, Kidnapped in Washington City in 1841, and Rescued in 1853, from a Cotton Plantation near the Red River in Louisiana* (New York: Miller, Orton & Mulligan, 1855), 260; Olmsted, *Cotton Kingdom,* 301.

37. Clara M. Winston to Patsy Morris, 16 July 1844, Guide to Dabney and Davis Papers, UVA; "Sippy" [Virginia Gordon] to Harriet E. Brown, 8 August 1849, Whitaker and Meade Family Papers, UNC; "Mary" to Mrs. James R. Aiken, 21 February 1851, Aiken Papers, Miscellaneous Manuscripts, USC; Abby M. Manly and T. M. Manly to B. Manly, 22 March 1853, Manly Papers, USC; Carney Diary, 8 June 1859, 25 May 1859, 14 April 1859, 15 July 1859, UNC. On men, see O. G. Murrell to John Murrell, 26 February 1850, Murrell Family Papers, UVA.

38. "Martha" to her mother, 1 January 1858, Scarborough Family Papers, DU; Marianne Gaillard to John L. Palmer, 8 August 1844, Palmer Family Papers, USC.

39. Elizabeth Blaetterman to "Victoria," 30 June 1860, Blaetterman Letter, UVA; G. F. Townes to J. A. Townes, 3 April 1834, Townes Family Papers, USC. See Rawick, ed., *American Slave, Alabama,* 6:58–59, for a plantation mistress who intervened to protect her slave women from an abusive overseer; see Herbert G. Gutman, *The Black Family in Slavery and Freedom, 1750–1925* (New York: Pantheon Books, 1976), 287–88, for a white man and woman who helped slaves write to family members.

40. Adelaide Crain to Caroline Gordon, 3 September 1857, Gordon and Hackett Family Papers, UNC; Ann E. Harris and James W. Harris to Sarah P. Hamilton, 29 April 1838, Benjamin Cudworth Yancey Papers, UNC.

41. Ann E. Harris and James W. Harris to Sarah P. Hamilton, 20 March 1837, Benjamin Cudworth Yancey Papers, UNC; Gayle Diary, 21 July 1828, Bayne and Gayle Family Papers, UNC; Yetman, *Peculiar Institution,* 13, 218.

42. Thomas Felix Hickerson, *Happy Valley: History and Genealogy* (Chapel Hill: by the author, 1940), 46; Addie and Carrie Dogan to Caroline Gordon, 14 June 1848, "Adelaide" [Stokes] to R. F. Hackett, 7 January 1846, Adelaide Crain to Caroline Gordon, 11 November 1852, Gordon and Hackett Family Papers, UNC.

Index

CPSIA information can be obtained
at www.ICGtesting.com
Printed in the USA
BVHW030335050121
596983BV00002B/12

9 780801 849640